The Solution of Partial Differential Equations by Finite Difference Approximations

Analysing the Relative Performance of Differing Numerical Finite Difference Schemes using Taylor Series Expansions

By

Lewis Hall MSc

Copyright © 2018 Lewis Hall

Copyright © 2018 Lewis and Lynn Aerospace Ltd

All Rights Reserved. No part of this publication may be reproduced, distributed, or transmitted in any form or by any electronic or mechanical means, including but not limited to; photocopying, recording, database storage, information and retrieval systems or other electronic or mechanical methods, without the express prior written permission of the publisher and the author of this book. Except in the case of brief quotations embodied in critical reviews, brief excerpts and references in academic papers, and certain other non-commercial uses permitted by copyright law.

For permission requests contact Lewis Hall or LLAerospace.com

This book is for entertainment, information and learning purposes. The publisher and author of this book are not responsible in any manner whatsoever for any adverse effects arising directly or indirectly as a result of the information provided in this book.

Visit **LLAerospace.com**

Twitter: **@LLAerospace**

Instagram: **@LLAerospace**

Merchandise Store: **https://shop.spreadshirt.co.uk/ll-aerospace**

About the Author

Lewis Hall attended the University of Salford studying Aeronautical Engineering at undergraduate level achieving a (BEng) First-Class Honours degree before advancing to study a postgraduate Master's of Science (MSc) degree in Aerospace Engineering achieving a Distinction. Were he specialised in Autonomous Predictive Non-linear Flight Control System Design, Flight Dynamics and Aerodynamics. Culminating in a thesis in which he designed (from conception to working prototype) an optimally performing, fully autonomous predictive Internal Model Control System for non-linear Stealth Fighter Aircraft, with full applicability to other Non-linear aircraft configurations.

Subsequently, he progressed into a career within the Aerospace industry, founded and operates the company Lewis and Lynn Aerospace, whilst writing books that distil complex engineering topics making them clear and comprehensive to further the knowledge and skills of the reader. In addition to his endeavours within Aerospace, Lewis has a profound interest in, hiking mountains, sailing, lifting heavy weights, Brazilian Ju-Jitsu and sports, particularly American Football and MMA.

Preface

This book is a comprehensive performance analysis of the Finite Difference Method for the solution of Partial Differential Equations. Providing an in-depth understanding of; Finite Difference Methods, their applications, theoretical basis, the full derivation of Taylor Series Expansions and the construction of a working Computational Domain Grid System. Furthermore, detailing and showing how to effectively employ the Finite Difference Method, through the implementation of Finite Difference Schemes, to obtain accurate, stable and consistent numerical solutions for Partial Differential Equations, which model a multitude of varying dynamic processes.

Moreover, it contains a detailed, thorough performance analysis investigation of three different Finite Difference Method schemes, when they are employed to obtain accurate numerical solutions for a fluid flow heat transfer process that is modelled by a first order Partial Differential Equation. These three schemes are the Forward-Time-Backwards-Space, Lax and Lax Wendroff Finite Difference Method schemes. Additionally, it explains the criteria that is required for optimal scheme stability, consistency and convergence.

Table of Contents

Chapter One: Introduction 1
   1.1: Performance Investigation Specification 5
      1.1.1: The Initial Conditions of the Finite Difference Grid System 7
      1.1.2: The Boundary Conditions 7
      1.1.3: Discretised Finite-Difference Scheme Equations 8
   1.2: The Objectives of the Investigation.. 9

Chapter Two: The Theory of Finite Difference Approximations 11
   2.1: Model Equation of the Linear Convection of Temperature 11
   2.2: The Finite Difference Method.. 13
      2.2.1: Discretisation 14
   2.3: Finite Difference Schemes 15
      2.3.1: The Forward-Time, Backwards-Space Scheme 17
      2.3.2: The Lax Scheme 18
      2.3.3: The Lax-Wendroff Scheme 18
   2.4: Convergence of Finite Difference Schemes................................. 19
      2.4.1: The Lax Equivalence Theorem 20
   2.5: The Stability of Finite Difference Schemes 21
   2.6: The Consistency of Finite Difference Schemes 21
      2.6.1: Taylor Series Expansions 22
   2.7: Finite Difference Scheme Errors 23
      2.7.1: Round-Off Error 23
      2.7.2: Truncation Error 23

Chapter Three: The Finite Difference Computational Grid System 25

    3.1: Developing the Discretised Finite Difference Computational Grid System 25

    3.2: Calculating the Numerical Solutions and Plotting Graphical Outputs 28

Chapter Four: Taylor Series Expansions 31

    4.1: Investigating the Consistency of a Finite Difference Scheme 31

    4.2: Taylor Series Expansion of the Temperature Variables.....................32

    4.3: Taylor Series Expansion of the Forward-Time, Backwards-Space Scheme 33

        4.3.1: Consistency Verification of the Forward-Time, Backwards-Space Scheme 34

        4.3.2: Truncation Error and Order of Accuracy of the FTBS Scheme...35

    4.4: Taylor Series Expansion of the Lax Scheme 35

        4.4.1: Consistency Verification of the Lax Scheme 38

        4.4.2: Truncation Error and Order of Accuracy of the Lax Scheme 38

    4.5: Taylor Series Expansion of the Lax-Wendroff Scheme......................40

        4.5.1: Consistency Verification of the Lax-Wendroff Scheme 41

        4.5.2: Truncation Error and Order of Accuracy of the Lax-Wendroff Scheme 42

Chapter Five: Numerical Solutions and Performance Analysis 43

    5.1: Forward-Time, Backwards-Space Scheme 43

        5.1.1: Numerical Solutions and Graphical Outputs................................43

        5.1.2: FTBS Scheme Performance Analysis 53

    5.2: Lax Scheme...57

        5.2.1: Numerical Solutions and Graphical Outputs................................57

        5.2.2: Lax Scheme Performance Analysis..66

    5.3: Lax-Wendroff Scheme 69

    5.3.1: Numerical Solutions and Graphical Outputs 69

    5.3.2: Lax-Wendroff Scheme Performance Analysis............................78

Chapter Six: Discussion..82

    6.1: Importance of Consistency..82

    6.2: Effects of Truncation Error 83

    6.3: Impact of Time-Step Size on Scheme Stability and Performance 84

    6.4: Discussion of Finite Difference Method Scheme Performance..........86

    6.5: Evaluation of Scheme Performance 87

    6.6: The Optimally Performing Finite Difference Scheme 88

Chapter Seven: Investigation Conclusion..89

    7.1: Sequential Investigation Suggestions...90

    7.2: Improvements to the Investigation...91

Chapter 8: References 92

List of Figures

Figure 2.0: A Visual Representation of a Temperature Profile Moving Linearly Downstream in a Fluid Flow……………………………………....12

Figure 2.1: A Discretised Finite Difference Method Grid System……………………………………………………………………….14

Figure 2.2: A Visual Representation of the Calculation Procedure of the Forward-Time, Backwards-Space Finite Difference Method ……………....17

Figure 3.0: The Structure of the Finite Difference Computational Grid System……………………………………………………………….…....26

Figure 5.0: Graphs showing the Convection of Temperature Downstream in a Fluid Flow for the FTBS Finite Difference Scheme at the 12 Computational Grid Points (Time Intervals of 0.01s) when $\Delta t = 0.01$ and $v = 0.05$…………………………………………………………………..45

Figure 5.1: Graphs showing the Convection of Temperature Downstream in a Fluid Flow for the FTBS Finite Difference Scheme at the 12 Computational Grid Points (Time Intervals of 0.1s) when $\Delta t = 0.1$ and $v = 0.5$……………………………………………………………………48

Figure 5.2: Graphs showing the Convection of Temperature Downstream in a Fluid Flow for the FTBS Finite Difference Scheme at the 12 Computational Grid Points (Time Intervals of 0.12s) when $\Delta t = 0.12$ and $v = 0.6$…………………………………………………………….......51

Figure 5.3: Graphs showing the Convection of Temperature Downstream in a Fluid Flow for the Lax Finite Difference Scheme at the 12 Computational Grid Points (Time Intervals of 0.01s) when $\Delta t = 0.01$ and $v = 0.05$...58

Figure 5.4: Graphs showing the Convection of Temperature Downstream in a Fluid Flow for the Lax Finite Difference Scheme at the 12 Computational Grid Points (Time Intervals of 0.1s) when $\Delta t = 0.1$ and $v = 0.5$..61

Figure 5.5: Graphs showing the Convection of Temperature Downstream in a Fluid Flow for the Lax Finite Difference Scheme at the 12 Computational Grid Points (Time Intervals of 0.12s) when $\Delta t = 0.12$ and $v = 0.6$ 64

Figure 5.6: Graphs showing the Convection of Temperature Downstream in a Fluid Flow for the Lax-Wendroff Finite Difference Scheme at the 12 Computational Grid Points (Time Intervals of 0.01s) when $\Delta t = 0.01$ and $v = 0.05$...70

Figure 5.7: Graphs showing the Convection of Temperature Downstream in a Fluid Flow for the Lax-Wendroff Finite Difference Scheme at the 12 Computational Grid Points (Time Intervals of 0.1s) when $\Delta t = 0.1$ and $v = 0.5$..73

Figure 5.8: Graphs showing the Convection of Temperature Downstream in a Fluid Flow for the Lax-Wendroff Finite Difference Scheme at the 12 Computational Grid Points (Time Intervals of 0.12s) when $\Delta t = 0.12$ and $v = 0.6$...76

List of Equations

1.0: The Partial Differential Equation for Linear Convection of Temperature in a Fluid Flow……………………………………………………………..…..5

1.1: The Discretised Finite Difference Equation for the Forward-Time, Backwards-Space Scheme…………………………………………………...8

1.2: The Discretised Finite Difference Equation for the Lax Scheme………..8

1.3: The Discretised Finite Difference Equation for the Lax-Wendroff Scheme……………………………………………………………………...8

1.4: Equation detailing the Finite Difference parameter v…………………...9

2.0: The Partial Differential Equation for the Linear Convection of Temperature in a Fluid Flow……………………………………...………..11

2.1: Taylor Series Expansion Model Definition……………………….....22

3.0: T_i^{n+1} Modified Discretised Forward-Time, Backward-Space Scheme Equation…………………………………………………………………..27

3.1: T_i^{n+1} Modified Discretised Lax Scheme Equation…………………..27

3.2: T_i^{n+1} Modified Discretised Lax-Wendroff Scheme Equation…………27

3.3: T_i^{n+1} Modified Discretised Forward-Time, Backward-Space Scheme Verification Equation……………………………………………...……..29

3.4: T_i^{n+1} Modified Discretised Lax Scheme Verification Equation….…...29

3.5: T_i^{n+1} Modified Discretised Lax-Wendroff Scheme Verification Equation……………………………………………………………..……29

4.0: T_i^{n+1} Variable Taylor Series Expansion………………………….32

4.1: T_{i+1}^{n} Variable Taylor Series Expansion……………………….....32

4.2: T_{i-1}^n Variable Taylor Series Expansion………………………….…...32

4.3: The Discretised Scheme Equation that is equal to zero for the FTBS Scheme…………………………………………………………………….…...33

4.4: The initial Taylor Series expansion for the FTBS Finite Difference Scheme…………………………………………………………………….……..33

4.5: The Rearranged Taylor Series Expansion for the FTBS Finite Difference Scheme……………………………………………………………………...…..34

4.6: The Final Taylor Series Expansion for the FTBS Finite Difference Scheme……………………………………………………………………..…...34

4.7: The FTBS Taylor Expansion as *Δx and Δt* tend to Zero………………..34

4.8: Truncation Error for the FTBS Finite Difference Scheme…………….35

4.9: The Discretised Scheme Equation that is equal to zero for the Lax Scheme……………………………………………………………….…..…...35

4.10: The initial Taylor Series expansion for the Lax Finite Difference Scheme……………………………………………………………………….…..36

4.11: The simplification step one of Taylor Series Expansion for the Lax Finite Difference Scheme…………………………………….……….....……..37

4.12: The simplification step two of Taylor Series Expansion for the Lax Finite Difference Scheme……………………………………………….……..37

4.13: The Final Taylor Series Expansion for the Lax Finite Difference Scheme………………………………………………………………….……...37

4.14: The LaxTaylor Expansion as *Δx and Δt* tend to Zero……………....…38

4.15: Truncation Error for the Lax Finite Difference Scheme………....…..38

4.16: The Discretised Scheme Equation that is equal to zero for the Lax-Wendroff Scheme……………………………………………….…….…..40

4.17: The initial Taylor Series expansion for the Lax-Wendroff Finite Difference Scheme…………………………………………….…..…......40

4.18: The Rearranged Taylor Series Expansion for the Lax-Wendroff Finite Difference Scheme…………………………………………..…...………41

4.19: The Final Taylor Series Expansion for the Lax-Wendroff Finite Difference Scheme…………………………………………………….........41

4.20: The Lax-Wendroff Taylor Expansion as Δx and Δt tend to Zero……..41

4.21: Truncation Error for the Lax-Wendroff Finite Difference Scheme.....42

Introduction

Numerical techniques have been employed as methods to solve both ordinary and partial differential equations for decades. Leonhard Euler was a pioneer of numerical analysis techniques in the 18th century when he conducted research into fluid dynamics and mathematical models, but his work primarily centred on solving ordinary differential equations. Numerical techniques to solve partial differential equations were not developed thoroughly until the 1930's, when a paper by Courant, Lewy, and Friedrichs introduced the Finite Difference method as a way of solving partial differential equations (Thomee, 2001)[1].

The theoretical basis of using Finite Differences to approximate solutions of partial derivatives was subsequently developed. However, it was not practically applicable until effective computer software programmes were operational and functional in the 1960's. Ever since, numerical techniques have been employed globally to solve both simple and complex partial differential equations using computer software, which perform the necessary calculations required to obtain solutions. These solutions are achieved by harnessing various Finite Difference schemes in discretised grid systems to approximate Finite Difference solutions for partial derivatives.

Finite Difference approximations are solutions for individual derivatives of partial differential equations that have been obtained from the application of the Finite Difference method. The Finite Difference method is a numerical technique that is often harnessed to model and ultimately obtain numerical solutions for a multitude of complex phenomena in fields such as mathematics, aerodynamics or fluid mechanics. Finite Difference methods enable the numerical solutions for partial differential equations to be acquired by the approximating derivatives through the utilisation of discretised equations in a Finite Difference grid system. These discretised equations approximate Finite Differences between grid points within that system or Finite Difference scheme.

Hence, Finite Difference schemes are methods of discretization that can be employed to produce accurate numerical solutions for processes and systems that are represented by partial differential equations. Finite Difference methods can provide a high level of accuracy for the numerical solutions. Consequently, this numerical technique has become the most dominant and commonly used process to obtain numerical solutions for partial differential equations (Christian Grossmann, 2007)[2].

Partial differential equations consist of derivatives of unknown multivariable functions that relate numerous different continuous independent variables together to describe a dynamic system (Bernoff, 2008)[3]. The diverse array of phenomena that partial differential equations can be used to describe is considerable, given their high versatility and tremendous usefulness in formalising such phenomena into an accessible form.

Consequently, partial differential equations are commonly used in areas such as thermodynamics, electrostatics and aerodynamics to describe and model heat transfer, electronic conduction and fluid flow problems, respectively. They can be applied to such a wide variety of systems because of their ability to contain first order, second order or even higher order derivatives, to describe the function of a multidimensional system. Hence,

Chapter 1   Introduction

they are highly versatile and valuable when employed to model multifaceted dynamic problems or systems.

As a result, they are critically important tools in scientific fields such as the aerospace and automotive industries. Were their extensive proliferation has enabled them to become the most commonly used and effective means of modelling physical problems such as the linear convection of temperature in a fluid flow. Although partial differential equations are commonly used in engineering, obtaining analytic numerical solutions for them is notoriously difficult. Therefore, various Finite Difference schemes have been developed and are used to solve them using continuous discretised mesh or finite grid systems.

The relative performance of these schemes and accuracy of the solutions obtained varies depending on; the type of partial differential being solved, the time step used, the schemes stability, consistency, and convergence. As well as the effect of the Truncation Errors. Thus, it is necessary to investigate the relative performance of different Finite Difference schemes in terms of their accuracy, consistency, stability, and convergence when solving a partial differential equation. So that the most suitable scheme can be defined to improve the accuracy of the numerical solutions for a given a process.

This book highlights an investigation and analysis of the relative performance of three different explicit Finite Difference schemes. Detailing the steps required to conduct a performance analysis of a Finite Difference scheme and, how to discern and analyse the results so that a comprehensive understanding of Finite Difference methods can be ascertained. Thus, enabling the most effective and applicable Finite Difference scheme for any type of system or problem that is described by partial differential equations, to be identified.

The relative performance of three different Finite Difference schemes is investigated in this book. Those being the Forward-Time, Backward-Space, Lax, and Lax-Wendroff, schemes. Their relative performance will be determined in terms of solving the linear convection of temperature partial differential equation and the schemes ability to provide accurate, stable numerical solutions.

The investigation is structured by a series of objectives that include analysing the stability of each scheme using various time step sizes and investigating the consistency of each scheme with the original partial differential equation using Taylor Series expansions. Through analysis of the results from these two investigations in addition to consulting the theoretical background associated with the performance of Finite Difference schemes, the relative performance of each scheme is determined. Furthermore, the most suitable Finite Difference scheme for solutions to linear partial differential equations will be established from the results of the investigation.

Chapter 1 — Introduction

1.1 Performance Investigation Specification

The partial differential equation that describes linear convection of temperature in a fluid flow, to which the various Finite Difference schemes will be applied to provide solutions for, is given by the following equation:

$$\frac{\partial T}{\partial t} + U\frac{\partial T}{\partial x} = 0 \qquad (1.0)$$

Where the temperature of the fluid is T and the convection velocity in the x-direction is U, which incidentally is constant for this linear convection system.

Numerical solutions for this partial differential equation are to be obtained to highlight the linear convection of temperature distribution in a fluid flow over time, through the application of three different Finite Difference schemes, which are;

- Forward-Time, Backward-Space
- Lax
- Lax-Wendroff

The relative performance of these different numerical Finite Difference schemes is to be investigated by analysing the stability, consistency, convergence of the schemes and the accuracy of the resultant numerical solutions. Effective investigation of the relative performance of each of the schemes can be achieved by developing a Finite Difference grid system containing twenty-one grid points.

Chapter 1   Introduction

For this investigation the grid system has a constant convection velocity of $U = 1.0$ and consists of twenty equally spaced intervals in the x-direction that are of the magnitude $\Delta x = 0.1$. Across which the temperature of the fluid flow (T) can be determined. The numerical solution for the temperature at each grid point and the temperature distribution across the whole system can then be obtained for various values of time-step (Δt) by implementing the discretised equation of each of the three different Finite Difference schemes into a Microsoft Excel spreadsheet to simulate the Finite Difference system.

In order to ensure suitability of the required Finite Difference grid system to enable solutions to be obtained for the partial differential equation describing the linear convection of temperature, it is necessary to define the initial conditions and boundary conditions of the system properly. Given that if excessive conditions are imposed there will be no solutions attainable, likewise, if too few conditions are specified, the numerical solutions for the equation will cease to be unique (Professor D.M. Causon, 2010)[4].

It is also appropriate when specifying the problem to describe each of the three-different discretised Finite Difference scheme equations that will be used to investigate relative performance and obtain numerical solutions for the partial differential equation for the linear convection of temperature in a fluid flow (equation 1.0).

1.1.1 The Initial Conditions of the Finite Difference Grid System

The initial conditions of the system are used as the foundations from which the relative performance of each of the Finite Difference schemes can be investigated and determined. The initial conditions highlight the temperature distribution in the fluid flow across the twenty-one different Finite grid points when time, $t = 0$.

The initial temperature distribution is;

$T_1 = 0.0$, $T_2 = 1.0$, $T_3 = 2.0$, $T_4 = 2.0$, $T_5 = 1.0$ and $T_i = 0.0$ for $i = 6$ to 21

1.1.2 The Boundary Conditions

To establish an accurate representation of the Finite Difference grid system the boundary conditions of the two end grid points T_1 and T_{21} respectively are required to be specified and are to remain fixed throughout the investigation for all values of time, t. Irrespective of the time step employed in the different Finite Difference schemes. As a result, the fixed boundary conditions of the Finite Difference grid system are defined as;

$T_1 = 0.0$ and $T_{21} = 0.0$ for all values of time, *t*.

1.1.3 Discretised Finite-Difference Scheme Equations

The discretised equations that describe each of the schemes that are to be investigated are given as;

Forward-Time, Backward-Space Scheme;

$$\frac{T_i^{n+1} - T_i^n}{\Delta t} + U\left(\frac{T_i^n - T_{i-1}^n}{\Delta x}\right) = 0 \qquad (1.1)$$

Lax Scheme;

$$\frac{T_i^{n+1} - \frac{T_{i+1}^n + T_{i-1}^n}{2}}{\Delta t} + U\left(\frac{T_{i+1}^n - T_{i-1}^n}{2\Delta x}\right) = 0 \qquad (1.2)$$

Lax-Wendroff Scheme;

$$T_i^{n+1} = T_i^n - U\Delta t\left(\frac{T_{i+1}^n - T_{i-1}^n}{2\Delta x}\right) + U^2\frac{\Delta t^2}{2}\left(\frac{T_{i+1}^n - 2T_i^n + T_{i-1}^n}{\Delta x^2}\right) \qquad (1.3)$$

These equations are employed to obtain numerical solutions for the temperature at the various grid points of space and time that make up a Finite Difference grid system.

1.2 The Objectives of the Investigation

The aim of this investigation is to determine the relative performance of three different numerical Finite Difference schemes, the Forward-Time Backward-Space, Lax and Lax-Wendroff schemes, in terms of the consistency, stability, convergence, and accuracy of the schemes when solving the linear convection of temperature in a fluid flow partial differential equation.

The successful fulfilment of the aim of the investigation to analyse the relative performance of the Forward-Time Backward-Space, Lax and Lax-Wendroff Finite-Difference schemes will be achieved by the completion of the following investigation objectives;

- For each of the individual Finite-Difference schemes that are to be investigated, tabulate results and produce graphical plots of the temperature of the fluid flow at each of the 21 Finite-Difference grid points over a period of twelve time-steps, for various values of the parameter v, were:

$$v = \frac{U \Delta t}{2 \Delta x} \qquad (1.4)$$

- Investigate the consistency of the Finite-Difference approximations with the original partial differential equation, that represents the linear convection of temperature in a fluid flow, for each of the three different schemes.

- Using the results obtained from the investigation, analyse the stability characteristics of each of the three Finite-Difference schemes.

- Evaluate the relative performance of the three Finite-Difference schemes, indicating the advantages and disadvantages of the application of each scheme when employed to obtain solutions for linear partial differential equations.

Upon completion of the investigation, comprehensive conclusions will be developed regarding the accuracy, stability, and consistency of the three-different Finite Difference schemes. Furthermore, based on these conclusions recommendations will be provided for the most suitable scheme that should be employed to obtain accurate, stable and consistent numerical solutions for linear partial differential equations.

Chapter 2 — Theory

The Theory of Finite Difference Approximations

This chapter of the book highlights and explains the theory that underpins Finite Difference approximations and theoretical background of this investigation into the relative performance of the FTBS, Lax and Lax-Wendroff Finite Difference schemes in terms of consistency, stability and accuracy of the schemes and numerical solutions obtained for the linear convection of temperature in a fluid flow.

2.1 Model Equation of the Linear Convection of Temperature

The model equation that is employed to represent the linear convection of temperature in a fluid flow is;

$$\frac{\partial T}{\partial t} + U \frac{\partial T}{\partial x} = 0 \qquad (2.0)$$

This hyperbolic equation is composed of first-order temporal and spatial derivatives and is employed to model the convection of temperature (heat) in a fluid flow with respect to time as it travels in the x-direction.

The spatial derivative ∂x represents the transport of temperature in the x-direction. Moreover, the temporal derivative ∂T represents the change in temperature as it travels downstream. Additionally, equation 2.0 highlights that the process of temperature convection downstream in a fluid flow is linear, since the differentials that describe the independent variables in the function are both first order.

Furthermore, equation 2.0 shows that convection of heat in a fluid flow and the accompanied temperature distribution in the x-direction is dependent on the rate of change of temperature with respect to change in time as well as

Chapter 2   Theory

convection velocity. Because the convection velocity determines the rate at which temperature distribution changes as it progresses downstream of the fluid flow in the x-direction.

As a result of the simplicity of this partial differential equation (equation 2.0), the exact solution is obtainable when the initial temperature conditions are defined, and the convection velocity is constant.

Therefore, given that the initial conditions of the equation are defined as;

$$f(x, t1 = 0) = F(x)$$

This is accurate for any point initial point in x. Thus, the exact solution of this partial differential equation occurs when the initial temperature profile moves consistently downstream without the values of the numerical solutions for temperature changing. This phenomenon is showcased in figure 2.0 (University of Alaska, 2004)[5].

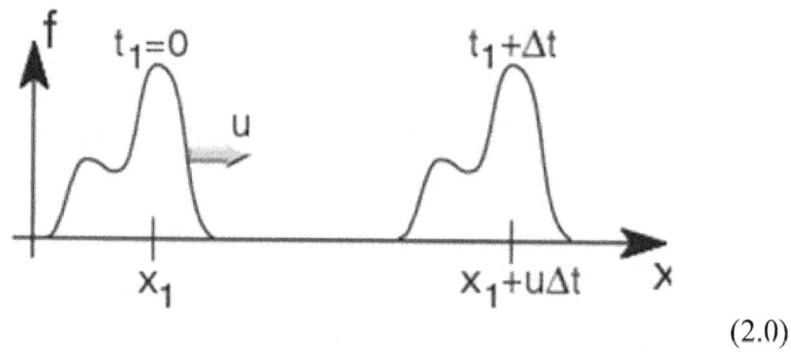

(2.0)

Figure 2.0: A Visual Representation of a Temperature Profile Moving Linearly Downstream in a Fluid Flow.

Consequently, the exact solution for the linear convection of temperature (equation 2.0) will be obtained when the numerical solutions from a Finite Difference scheme does not change from the initial temperature conditions when progressing through the fluid flow in the x-direction. Thus, preserving the initial temperature profile.

2.2 The Finite Difference Method

The Finite Difference Method is the most commonly employed numerical technique used to provide approximate numerical solutions to partial differential equations that cannot be solved analytically to provide definable exact solutions. These approximate numerical solutions require the partial differential to be discretised so that the independent variables in the equation can be defined at grid points in a Finite Difference grid system (also known as a mesh system). This enables discrete Finite Difference approximations of the continuous derivatives to be obtained from those grid points, therefore enabling a solution for the dependent variable to be obtained.

The fundamental basis of the Finite Difference Method is Taylor's Theorem which uses neighbouring values in a Finite Difference grid to approximate the derivatives at any point in a system. This is achieved through the application of a Taylor series expansion to an equation describing a Finite Difference scheme (York, 2009)[6].

2.2.1 Discretisation

Discretisation is the process of converting a continuous function into finite discreet points in time and space that are definable on a grid system so that numerical solutions of the continuous function can be approximated at those points.

The Finite Difference Method is an example of discretisation because it enables a partial differential to be expressed as an algebraic equation. This differential can then be solved in a spatial and temporal computational domain grid system to obtain numerical solutions from previously calculated results. An example of the computational grid system that is used in this investigation is shown in figure 2.1 (Johnston, 2014)[7].

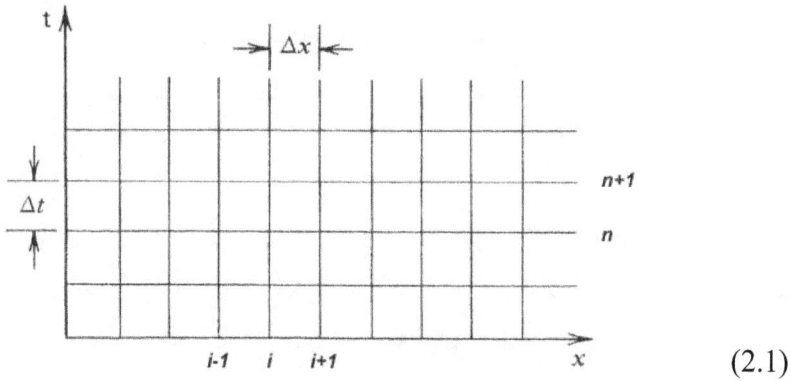

(2.1)

Figure 2.1: A Discretised Finite Difference Method Grid System

2.3 Finite Difference Schemes

The Finite Difference Method can be used to numerically solve all types of partial differential equations. However, to ensure that effective and accurate solutions can be obtained, different schemes are required to be applied dependant on the nature of the partial differential equation.

If the equation is parabolic then schemes such as the Crank-Nicholson or Forward-Time, Centred-Space should be applied. Whereas if the model equation is hyperbolic, (such as the linear convection of temperature in a fluid flow equation used in this investigation), then schemes such as the Forward-Time, Backwards-Space, Lax and Lax-Wendroff should be applied to obtain numerical solutions. Implementing these Finite Difference schemes to a developed Finite Difference grid computational domain system ensures that numerical solutions for a partial differential equation are obtainable for all values of time t.

Irrespective of the type of Finite Difference scheme, there are three distinct properties that need to be assessed in order to determine the relative performance of a scheme and to enable accurate numerical solutions to be obtained for a partial differential equation.

These are;

- Consistency
- Stability
- Convergence

Furthermore, because of the continuous nature of partial differential equations, and the use of a computational domain to calculate solutions from previous results in a Finite Difference grid system, there are two distinctive sources of error that commonly effect Finite Difference schemes. These errors affect the resultant numerical solution by reducing the accuracy, stability, and convergence of a scheme. These errors can be particularly impactful and corrode the effectiveness of a Finite Difference scheme if allowed to compound throughout numerical solutions.

The sources of error in a Finite Difference scheme are;

- Truncation Error

- Round-off Error

As a result of establishing the properties of Finite Difference schemes and possible sources of error in the solutions, it is appropriate to describe the individual discretised Finite Difference schemes for which the relative performance is being assessed in this investigation.

Chapter 2   Theory

2.3.1 The Forward-Time, Backwards-Space Scheme

The Forward-Time, Backwards-Space or FTBS scheme is an explicit numerical technique that can be applied to hyperbolic partial differential equations, to provide numerical solutions. The procedure used by this scheme to calculate solutions is a backwards difference method which works by calculating successive numerical solutions from previous results in the grid system. This process is highlighted in figure 2.2 (MIT, 2010)[8].

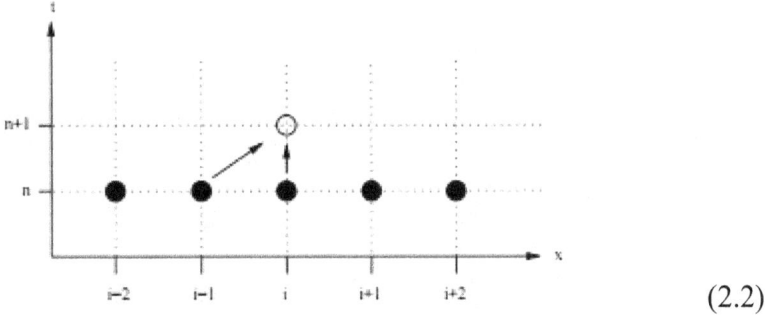

(2.2)

Figure 2.2: A Visual Representation of the Calculation Procedure of the Forward-Time, Backwards-Space Finite Difference Method

Furthermore, the FTBS scheme is conditionally stable depending on the time step size and is described by equation 1.1:

$$\frac{T_i^{n+1} - T_i^n}{\Delta t} + U\left(\frac{T_i^n - T_{i-1}^n}{\Delta x}\right) = 0 \qquad (1.1)$$

2.3.2 The Lax Scheme

The Lax Finite Difference scheme is a single step explicit scheme that is conditionally stable when used to solve linear partial differential equations. The condition of stability of this scheme is dependent on the time step size used in the Finite Difference discretised grid system. Moreover, the numerical solutions obtained from the Lax scheme are subject to excessive damping, therefore reducing its suitability and accuracy when solving linear partial differential equations.

The scheme is described by equation 1.2:

$$\frac{T_i^{n+1} - \frac{T_{i+1}^n + T_{i-1}^n}{2}}{\Delta t} + U\left(\frac{T_{i+1}^n - T_{i-1}^n}{2\Delta x}\right) = 0 \qquad (1.2)$$

2.3.3 The Lax-Wendroff Scheme

The Lax-Wendroff Finite Difference scheme is a second order accurate explicit Finite Difference scheme that is conditionally stable depending on the size of the time-step implemented in the Finite Difference grid system.

The scheme is described by equation 1.3:

$$T_i^{n+1} = T_i^n - U\Delta t\left(\frac{T_{i+1}^n - T_{i-1}^n}{2\Delta x}\right) + U^2 \frac{\Delta t^2}{2}\left(\frac{T_{i+1}^n - 2T_i^n + T_{i-1}^n}{\Delta x^2}\right) \qquad (1.3)$$

2.4 Convergence of Finite Difference Schemes

Convergence of a Finite Difference scheme is imperative for the scheme to be useful, and for it provide accurate numerical solutions for a partial differential equation. Since, only a convergent scheme can accurately provide solutions for a partial differential equation.

A Finite Difference scheme that is employed to provide numerical solutions for a partial differential equation can be described as convergent at any point in Finite Difference grid system (i.e. point (i,n)) if, the discretised equation of the scheme tends to the exact solution of the partial differential equation as the pointwise error at that point (i,n) tends to zero, in addition to the spatial and time variables (Δx and Δt) tending to zero (Professor D.M. Causon, 2010)[4].

Therefore;

$$e_i^n \to 0 \text{ as } \Delta t \text{ and } \Delta x \to 0 \text{ at } (i,n)$$

Convergence is extremely difficult to theoretically prove for many Finite Difference schemes. Nevertheless, for a partial differential that represents a linear process, such as linear convection of temperature, the Lax Equivalence Theorem can be applied to determine the properties that are required for a convergent Finite Difference scheme.

2.4.1 The Lax Equivalence Theorem

The Lax Equivalence Theorem states (Trefethen, 1994)[9];

"for a linear Finite Difference scheme that is considered consistent with the original partial differential equation when the independent variables tend to zero, the characteristic of a scheme that is necessary for convergence is stability"

Therefore, for Finite Difference numerical techniques, the relationship between the properties of a Finite Difference scheme show convergence as a characteristic of a consistent stable scheme.

Hence (Stanford, 2008)[10];

$$Consistency + Stability \Leftrightarrow Convergence$$

Consequently, for linear numerical methods (Finite Difference schemes) that are with consistent the Lax Equivalence theorem, the relationship becomes;

$$Stability \Leftrightarrow Convergence$$

As a result, a Finite Difference scheme that is proved to satisfy the consistency condition (i.e. is consistent) only has one criterion for convergence. The necessary condition for convergence is the stability of the scheme (J.H. Ferziger, 2002)[11]. Therefore, a linear Finite Difference scheme will be convergent if it is stable.

2.5 The Stability of Finite Difference Schemes

A Finite Difference scheme is quantified as stable if it does not magnify errors, such as Truncation Error, that are present in the numerical solutions. Since error magnification would cause the solutions to be inaccurate. Increasingly so, given the compounding nature of Finite Difference schemes. Stability of a Finite Difference scheme is required for it to produce accurate solutions.

The stability of a Finite Difference scheme is dependent on the impact and severity of the errors present within the numerical solutions obtained from a discretised Finite Difference grid system. A Finite Difference scheme will inherently remain stable if the round-off or Truncation Errors present within the numerical solutions produce only small perturbations throughout the grid system. Whilst, not excessively impacting upon the accuracy of the solutions (Urroz, 2004)[12]. If the errors in a Finite Difference scheme become unbounded and increase rapidly throughout the numerical solutions, they will continually increase and cause the system to become unstable.

2.6 The Consistency of Finite Difference Schemes

The consistency of a Finite Difference scheme determines how accurately the discretised scheme can model the original partial differential equation to obtain accurate and valid numerical solutions.

A Finite Difference scheme is considered to be consistent when the Taylor series expansion of the initial discretised equation reproduces the original form of the initial partial differential equation when the temporal (Δt) and spatial (Δx) step variables tend to zero. Verification of the consistency of a Finite Difference scheme requires a Taylor series expansion of the original scheme equation (equations 1.1, 1.2 and 1.3 for the schemes in this investigation) to be conducted to determine whether a scheme is consistent or inconsistent.

2.6.1 Taylor Series Expansions

A Taylor Series expansion of a discretised Finite Difference scheme equation is the procedure employed to examine the consistency of a Finite Difference scheme that is modelling a dynamic phenomenon represented by a partial differential equation. This is due to the Taylor Series enabling the sum of the derivatives of a function at a defined point to be determined through approximation. Taylor Series expansions are the direct application of Taylors Theorem that was developed by Brook Taylor in the 18th century, as a way of quantifying errors as estimations in approximated numerical solutions.

A Taylor series expansion is defined by equation 2.1 for a function *f(x)*:

$$f(x) = \sum_{n=0}^{\infty} \frac{f^{(n)} x_0}{n!} (x - x_0)^n \qquad (2.1)$$

Subsequently, this equation can be expanded to enable approximations of the partial derivatives of a system to be calculated. The complete Taylor series expansions for the FTBS, Lax and Lax-Wendroff schemes can be found in the Investigation Methodology chapter of this book.

2.7 Finite Difference Scheme Errors

There are two prominent types of error that occur when using a computational grid system to calculate numerical solutions of partial differential equations using Finite Difference schemes, they are Round-off Error and Truncation Error.

2.7.1 Round-Off Error

Round-off errors are present within the numerical solutions obtained from a Finite Difference scheme because of the finite arithmetic used by computers to calculate results or numerical solutions. Therefore, those calculations will only be conducted using a finite set of digits, i.e. 8 digits. Hence, the solutions will not be exact. Round-off errors do not directly affect the stability of a scheme or numerical solutions because they do not grow to become unbounded in continuous computation. Thus, they cannot cause the solutions to diverge from initial conditions. Nevertheless, they can have a significant impact on the accuracy of numerical solutions obtained from a Finite Difference computational domain system.

2.7.2 Truncation Error

The Truncation Error in a Finite Difference scheme is a considerably important factor in the stability of a scheme and accuracy of the numerical solutions it provides. Given that the error can grow unboundedly within a Finite Difference grid system if the time step size is too large.

Truncation Error is defined as the difference between the numerical scheme and the original differential equation (Hirsch, 2007)[13]. The Truncation Error in a scheme is the additional terms of the Taylor Series expansion when compared to the original discretised equation of the scheme. Furthermore, Truncation Error defines the order of accuracy of a Finite Difference scheme in space and time.

As a consequence of developing the theoretical background for this performance analysis investigation, the methodology and performance analysis work can be conducted. Since, the requirements for stability, consistency, convergence and accuracy of a Finite Difference scheme have been determined, in addition to providing explanations of the possible sources of error.

The Finite Difference Computational Grid System

This chapter details the development of a computational domain grid system using a series of Microsoft Excel spreadsheets to model the effects of varying time step size Δt (which subsequently changed the parameter v, given that the other variables that affected this parameter were constant) on the solutions for each of the Finite Difference schemes.

3.1 Developing the Discretised Finite Difference Computational Grid System

The development of the Finite Difference computational domain grid using Excel spreadsheets begins by constructing a Finite Difference grid system that consisted of twenty-one equally spaced grid points in the x-direction and twelve time-step intervals.

This is to enable the numerical solutions of the temperature distribution in the fluid flow to be obtained and plotted graphically for twelve time intervals. Subsequently, inserting the defined initial conditions and boundary conditions into this grid system enables the basis of the Finite Difference grid system to be finalised so that the three different schemes could be applied to it by implementing each of the Finite Difference equations into the grid system. Figure 3.0 highlights the structure of the embryonic Finite Difference computational grid system developed for this investigation before any of the equations are applied to it.

Chapter 3   Grid System

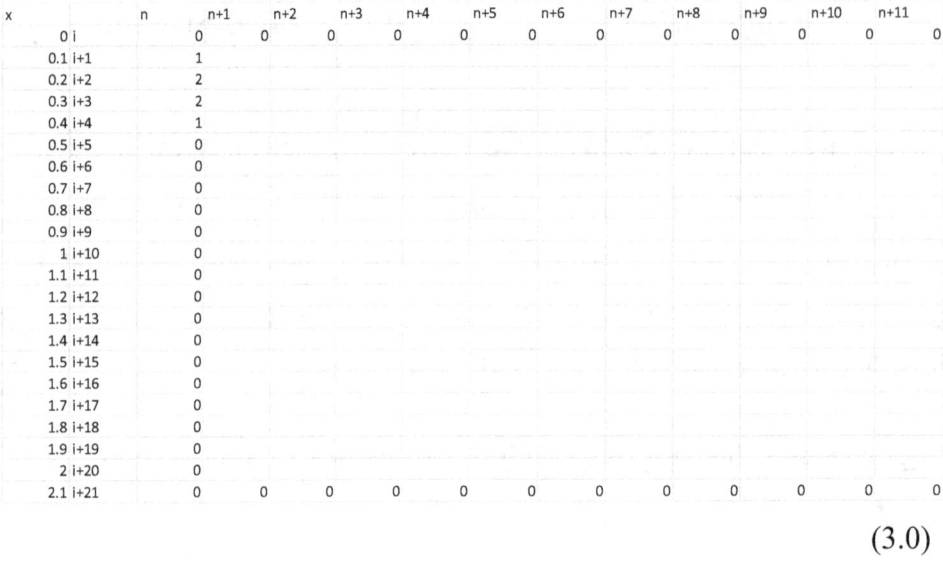

(3.0)

Figure 3.0: The Structure of the Finite Difference Computational Grid System

After completing the initial development of the embryonic Finite Difference computational domain grid system, the next phase of the investigation procedure is to apply the three different schemes to the system (figure 2.0). Via the implementation of the three unique scheme equations. Thus, the discretised Finite Difference scheme equations are added to the Finite Difference grid system, in their own individual spreadsheets.

To enable the equations to be inputted into the computational grid, the discretised Finite Difference equations required some substation and algebraic manipulation to ensure that T_i^{n+1} was the subject of each equation. Therefore, enabling the temperature solution to be obtained for each additional continuous time step in each scheme.

As a result of the T_i^{n+1} modification each of the discretised Finite Difference scheme model equations become;

$T_i^{\,n+1}$ Modified Discretised Forward-Time, Backward-Space Scheme Equation;

$$T_i^{n+1} = T_i^n - U\Delta t \left(\frac{T_i^n - T_{i-1}^n}{\Delta x} \right) \qquad (3.0)$$

$T_i^{\,n+1}$ Modified Discretised Lax Scheme Equation;

$$T_i^{n+1} = \frac{T_{i+1}^n + T_{i-1}^n}{2} - U\Delta t \left(\frac{T_{i+1}^n - T_{i-1}^n}{2\Delta x} \right) \qquad (3.1)$$

$T_i^{\,n+1}$ Modified Discretised Lax-Wendroff Scheme Equation;

$$T_i^{n+1} = T_i^n - U\Delta t \left(\frac{T_{i+1}^n - T_{i-1}^n}{2\Delta x} \right) + U^2 \frac{\Delta t^2}{2} \left(\frac{T_{i+1}^n - 2T_i^n + T_{i-1}^n}{\Delta x^2} \right) \qquad (3.2)$$

Now that the equations are in this form, implementation into the Finite Difference computational grid system can occur to form the final Finite Difference grid system for each of the three schemes. Enabling numerical solutions for the temperature in the fluid flow at each point in the Finite Difference grid system across each of the twelve time intervals to be obtained.

The next phase of the investigation is to specify and define the values of time-step, Δt, that are used to vary the parameter v and subsequently calculate the values of the parameter v using equation 1.4 that would be used within the Finite Difference computational system.

Three different values of v is sufficient to constitute enough variation of the results. Noting that, if the time steps are considerably different the solutions may become unstable. Particularly if a time-step much greater than 0.1 is employed.

The values of Δt and the corresponding value of v that are used in the investigation for each Finite Difference scheme system are defined in table 3.0;

Δt	v
0.01	0.05
0.1	0.5
0.12	0.6

Table 3.0: The values of Δt and v that are employed in the investigation.

3.2 Calculating the Numerical Solutions and Plotting Graphical Outputs

Defining the different time step values that are used to obtain each set of numerical solutions for the temperature in the fluid flow at twelve different time intervals, permits the Finite Difference equations for each scheme to be implemented into the Finite Difference computational grid system (Excel spreadsheet).

To verify that the spreadsheet works correctly it is critically important to check that the results obtained are correct. This is achieved by using the initial conditions of the fluid flow (temperature (i.e. temperature at n), a time step of $\Delta t = 0.01$, $\Delta x = 0.1$ and a constant convection velocity of 1.0) to conduct sample calculations for each of the three schemes to obtain solutions for the temperature at the point i+1, n+1.

Hence;

Forward-Time, Backward-Space Scheme;

$$T_i^{n+1} = 1 - (1.0 \times 0.01)\left(\frac{1-0}{0.1}\right) = 0.9 \qquad (3.3)$$

Lax Scheme;

$$T_i^{n+1} = \frac{2+0}{2} - (1.0 \times 0.01)\left(\frac{2-0}{2(0.1)}\right) = 0.9 \qquad (3.4)$$

Lax-Wendroff Scheme;

$$T_i^{n+1} = 1 - (1.0 \times 0.01)\left(\frac{2-0}{2(0.1)}\right) +$$

$$1^2 \frac{0.01^2}{2}\left(\frac{2 - 2(1) + 0}{0.1^2}\right) = 0.9 \qquad (3.5)$$

Observing and comparing the hand calculation results with the Finite Difference method computational grid system results (contained in the Numerical Solutions and Performance Analysis chapter) shows that the sample hand calculations for each Finite Difference scheme adhere to the results from the FDM computational grid system this point on the Finite Difference grid. Verifying that each of the equations in the Finite Difference system operates as intended.

Completing the verification calculations enables the development of the computational grid system to progress by using the cells that contain correct calculated results to determine the remaining numerical solutions at each of the grid points (this is possible by making use of Excels ability to calculate continuous functions from previous results). Therefore, creating a Finite Difference computational grid system.

Subsequently, once all of the numerical solutions have been calculated and verified, the creation of the graphical plots for the temperature distribution across each of the twenty-one grid points at each of the twelve independent time steps can occur (by using the graphing tool in Excel). The graphical plots and tabulated numerical solutions of the temperature and heat transfer in the fluid flow that were obtained during the performance analysis investigation can be observed in the Numerical Solutions and Performance Analysis chapter.

Taylor Series Expansions

This chapter details the Consistency verification of the three different Finite Difference Schemes and the corresponding Taylor Series Expansions, which is an integral part of a performance analysis investigation. Given that, scheme convergence is dependent on consistency with the initial Partial Differential Equation as determined by the Lax Equivalence Theorem. Additionally, deriving the Taylor Series expansions for the three Finite Difference Schemes, enables their Truncation Error and Order of Accuracy to be determined as well.

4.1 Investigating the Consistency of a Finite Difference Scheme

As discussed in The Theory of Finite Difference Approximations chapter, the consistency of a Finite Difference scheme requires the derivation of its corresponding Taylor Series expansion from the discretised equation of the scheme. This is performed to determine if the expansion returns to the original partial differential equation in terms of the point i, n when Δt and Δx tend to zero. If the equation does return to the form of the original partial differential the scheme is consistent. Whereas, if it does not the Finite Difference scheme is inconsistent.

Therefore, to verify the consistency, derivations of the Taylor Series expansions for the FTBS, Lax and Lax-Wendroff schemes are required. Enabling a consistency check to be performed.

Chapter 4

4.2 Taylor Series Expansion of the Temperature Variables

In order to commence with each of the Taylor series expansions efficiently it was deemed appropriate to initially derive the Taylor Series expansions of the temperature variables $T_i^{n+1}, T_{i+1}^n, T_{i-1}^n$.

$$T_i^{n+1} = \left(T_i^n + \Delta t \left.\frac{\partial T}{\partial t}\right|_i^n + \frac{\Delta t^2}{2!} \left.\frac{\partial^2 T}{\partial t^2}\right|_i^n + \frac{\Delta t^3}{3!} \left.\frac{\partial^3 T}{\partial t^3}\right|_i^n + \cdots \right) \quad (4.0)$$

$$T_{i+1}^n = \left(T_i^n + \Delta x \left.\frac{\partial T}{\partial x}\right|_i^n + \frac{\Delta x^2}{2!} \left.\frac{\partial^2 T}{\partial x^2}\right|_i^n + \frac{\Delta x^3}{3!} \left.\frac{\partial^3 T}{\partial x^3}\right|_i^n + \cdots \right) \quad (4.1)$$

$$T_{i-1}^n = \left(T_i^n - \Delta x \left.\frac{\partial T}{\partial x}\right|_i^n + \frac{\Delta x^2}{2!} \left.\frac{\partial^2 T}{\partial x^2}\right|_i^n - \frac{\Delta x^3}{3!} \left.\frac{\partial^3 T}{\partial x^3}\right|_i^n + \cdots \right) \quad (4.2)$$

4.3 Taylor Series Expansion of the Forward-Time, Backwards-Space Scheme

To commence with a Taylor Series expansion of a discretised Finite Difference equation, it is necessary to ensure that the Finite Difference scheme equation (equation 1.1) is equal to zero.

Therefore, the FTBS scheme is;

$$\frac{T_i^{n+1} - T_i^n}{\Delta t} + U\left(\frac{T_i^n - T_{i-1}^n}{\Delta x}\right) = 0 \qquad (4.3)$$

The previously derived the Taylor series expansions for each of different temperature variables (equations 4.0, 4.1 and 4.2), can be inputted into this equation to form the initial Taylor Series expansion for the FTBS scheme.

The initial Taylor Series expansion for the FTBS Finite Difference Scheme is;

$$\frac{1}{\Delta t}\left(T_i^n + \Delta t \left.\frac{\partial T}{\partial t}\right|_i^n + \frac{\Delta t^2}{2!}\left.\frac{\partial^2 T}{\partial t^2}\right|_i^n + \frac{\Delta t^3}{3!}\left.\frac{\partial^3 T}{\partial t^3}\right|_i^n + \cdots - T_i^n\right)$$

$$+ \frac{U}{\Delta x}\left(T_i^n - \left(T_i^n - \Delta x \left.\frac{\partial T}{\partial x}\right|_i^n + \frac{\Delta x^2}{2!}\left.\frac{\partial^2 T}{\partial x^2}\right|_i^n - \frac{\Delta x^3}{3!}\left.\frac{\partial^3 T}{\partial x^3}\right|_i^n + \cdots\right)\right)$$

$$= 0 \qquad (4.4)$$

Subsequent rearrangement and simplification of the Taylor Series expansion of the FTBS discretised equation gives;

$$\left(\left.\frac{\partial T}{\partial t}\right|_i^n + \frac{\Delta t}{2!}\left.\frac{\partial^2 T}{\partial t^2}\right|_i^n + \frac{\Delta t^2}{3!}\left.\frac{\partial^3 T}{\partial t^3}\right|_i^n\right) + \left(\left.U\frac{\partial T}{\partial x}\right|_i^n - U\frac{\Delta x}{2!}\left.\frac{\partial^2 T}{\partial x^2}\right|_i^n + U\frac{\Delta x^2}{3!}\left.\frac{\partial^3 T}{\partial x^3}\right|_i^n\right)$$
$$= 0 \qquad (4.5)$$

Further rearrangement of equation 4.5, provides the final Taylor Series expansion of the FTBS scheme, which is;

$$\left.\frac{\partial T}{\partial t}\right|_i^n + U\left.\frac{\partial T}{\partial x}\right|_i^n + \frac{\Delta t}{2!}\left.\frac{\partial^2 T}{\partial t^2}\right|_i^n - U\frac{\Delta x}{2!}\left.\frac{\partial^2 T}{\partial x^2}\right|_i^n + \frac{\Delta t^2}{3!}\left.\frac{\partial^3 T}{\partial t^3}\right|_i^n + U\frac{\Delta x^2}{3!}\left.\frac{\partial^3 T}{\partial x^3}\right|_i^n$$
$$= 0 \qquad (4.6)$$

Obtaining the full Taylor Series expansion of the Forward-Time Backwards-Space Finite Difference scheme enables the consistency of the scheme to be investigated.

4.3.1 Consistency Verification of the Forward-Time, Backwards-Space Scheme

To check the consistency of a Finite Difference scheme, that is described by a discretised equation, the spatial and time dependent variables of the Taylor series expansion are required to tend to zero.

Therefore, as *Δx* tends to zero and *Δt* tends to zero the Forward-Time, Backwards-Space Taylor Expansion equation (equation 4.6) becomes:

$$\left.\frac{\partial T}{\partial t}\right|_i^n + U\left.\frac{\partial T}{\partial x}\right|_i^n = 0 \qquad (4.7)$$

It can be observed by comparing equations 2.0 and 4.7, that when Δx and Δt tend to zero, the Taylor series expansion of the discretised FTBS equation is the original partial differential equation for the linear convection of temperature in a fluid flow evaluated at point i, n.

Hence, the Forward-Time, Backwards-Space Finite Difference scheme is **consistent**. Thus, producing consistent numerical solutions for the linear convection of temperature in a fluid flow.

4.3.2 Truncation Error and Order of Accuracy of the FTBS Scheme

The Taylor Series Expansion of the FTBS equation enables the Truncation Error of this scheme to be quantified by the following Truncation Error terms (were T.E. is the Truncation Error):

$$T.E. = \frac{\Delta t}{2!}\frac{\partial^2 T}{\partial t^2}\bigg|_i^n - U\frac{\Delta x}{2!}\frac{\partial^2 T}{\partial x^2}\bigg|_i^n + \frac{\Delta t^2}{3!}\frac{\partial^3 T}{\partial t^3}\bigg|_i^n + U\frac{\Delta x^2}{3!}\frac{\partial^3 T}{\partial x^3}\bigg|_i^n \quad (4.8)$$

Therefore, the leading Truncation Error terms in time and space for the FTBS scheme are of order Δt and Δx respectively. As a result the Forward-Time, Backward-Space Scheme is first order accurate in time and space.

4.4 Taylor Series Expansion of the Lax Scheme

To perform the Taylor Series expansion of the Lax scheme it is necessary to make the discretised Finite Difference equation (equation 1.2) that described the Lax scheme equal to zero.

$$\frac{T_i^{n+1} - \frac{T_{i+1}^n + T_{i-1}^n}{2}}{\Delta t} + U\left(\frac{T_{i+1}^n - T_{i-1}^n}{2\Delta x}\right) = 0 \quad (4.9)$$

The previously derived Taylor Series expansions for the temperature variables (equations 4.0, 4.1 and 4.2) can then be inputted into equation 4.9 to derive the initial Taylor Series expansion for the Lax scheme.

The initial Taylor Series expansion for the Lax Scheme is;

$$\frac{1}{\Delta t}\left(\left(T_i^n + \Delta t \left.\frac{\partial T}{\partial t}\right|_i^n + \frac{\Delta t^2}{2!}\left.\frac{\partial^2 T}{\partial t^2}\right|_i^n + \frac{\Delta t^3}{3!}\left.\frac{\partial^3 T}{\partial t^3}\right|_i^n + \cdots\right)\right.$$

$$-\frac{1}{2}\left(\left(T_i^n + \Delta x \left.\frac{\partial T}{\partial x}\right|_i^n + \frac{\Delta x^2}{2!}\left.\frac{\partial^2 T}{\partial x^2}\right|_i^n + \frac{\Delta x^3}{3!}\left.\frac{\partial^3 T}{\partial x^3}\right|_i^n + \cdots\right)\right.$$

$$\left.\left.+\left(T_i^n - \Delta x \left.\frac{\partial T}{\partial x}\right|_i^n + \frac{\Delta x^2}{2!}\left.\frac{\partial^2 T}{\partial x^2}\right|_i^n - \frac{\Delta x^3}{3!}\left.\frac{\partial^3 T}{\partial x^3}\right|_i^n + \cdots\right)\right)\right)$$

$$+\frac{U}{2\Delta x}\left(\left(T_i^n + \Delta x \left.\frac{\partial T}{\partial x}\right|_i^n + \frac{\Delta x^2}{2!}\left.\frac{\partial^2 T}{\partial x^2}\right|_i^n + \frac{\Delta x^3}{3!}\left.\frac{\partial^3 T}{\partial x^3}\right|_i^n + \cdots\right)\right.$$

$$\left.-\left(T_i^n - \Delta x \left.\frac{\partial T}{\partial x}\right|_i^n + \frac{\Delta x^2}{2!}\left.\frac{\partial^2 T}{\partial x^2}\right|_i^n - \frac{\Delta x^3}{3!}\left.\frac{\partial^3 T}{\partial x^3}\right|_i^n + \cdots\right)\right)$$

$$= 0 \qquad (4.10)$$

This initial Taylor Series expansion can be subsequently rearranged and simplified in two iterative steps to enable the final Taylor Series expansion for the Lax scheme to be obtained.

Simplification Step One:

$$\frac{1}{\Delta t}\left(\left(T_i^n + \Delta t\left.\frac{\partial T}{\partial t}\right|_i^n + \frac{\Delta t^2}{2!}\left.\frac{\partial^2 T}{\partial t^2}\right|_i^n + \frac{\Delta t^3}{3!}\left.\frac{\partial^3 T}{\partial t^3}\right|_i^n\right) - \left(T_i^n + \frac{\Delta x^2}{2!}\left.\frac{\partial^2 T}{\partial x^2}\right|_i^n\right)\right)$$

$$+ \frac{U}{2\Delta x}\left(2\Delta x\left.\frac{\partial T}{\partial x}\right|_i^n + \frac{2\Delta x^3}{3!}\left.\frac{\partial^3 T}{\partial x^3}\right|_i^n\right)$$

$$= 0 \tag{4.11}$$

Simplification Step Two:

$$\frac{1}{\Delta t}\left(\Delta t\left.\frac{\partial T}{\partial t}\right|_i^n + \frac{\Delta t^2}{2!}\left.\frac{\partial^2 T}{\partial t^2}\right|_i^n + \frac{\Delta t^3}{3!}\left.\frac{\partial^3 T}{\partial t^3}\right|_i^n - \frac{\Delta x^2}{2!}\left.\frac{\partial^2 T}{\partial x^2}\right|_i^n\right)$$

$$+ \frac{U}{2\Delta x}\left(2\Delta x\left.\frac{\partial T}{\partial x}\right|_i^n + \frac{2\Delta x^3}{3!}\left.\frac{\partial^3 T}{\partial x^3}\right|_i^n\right) = 0 \tag{4.12}$$

Further substitution of equation 4.12 renders the final Taylor Series expansion of the Lax scheme as;

$$\left.\frac{\partial T}{\partial t}\right|_i^n + U\left.\frac{\partial T}{\partial x}\right|_i^n + \frac{\Delta t}{2!}\left.\frac{\partial^2 T}{\partial t^2}\right|_i^n + \frac{\Delta t^2}{3!}\left.\frac{\partial^3 T}{\partial t^3}\right|_i^n - \frac{\Delta x^2}{2\Delta t}\left.\frac{\partial^2 T}{\partial x^2}\right|_i^n + U\frac{\Delta x^2}{3!}\left.\frac{\partial^3 T}{\partial x^3}\right|_i^n$$

$$= 0 \tag{4.13}$$

As a consequence of determining the final form of the Lax scheme Taylor Series expansion it enables the consistency verification of the Lax Finite Difference Scheme to be conducted as part of the performance analysis investigation.

4.4.1 Consistency Verification of the Lax Scheme

The Taylor Series expansion of the Lax Finite Difference scheme (equation 4.13) as Δx and Δt tends to zero is:

$$\left.\frac{\partial T}{\partial t}\right|_i^n + U\left.\frac{\partial T}{\partial x}\right|_i^n - \frac{\Delta x^2}{2\Delta t}\left.\frac{\partial^2 T}{\partial x^2}\right|_i^n = 0 \qquad (4.14)$$

The third term of equation 4.14 is indeterminate, given that it contains a divisible term (Δt) that would be equal to zero when both Δx and Δt tend to zero. As a result, this equation is not equivalent to the original partial differential equation (equation 2.0), therefore the discretised Lax scheme Finite Difference equation is **inconsistent**. Consequently, it is not useful for the solution of linear convection of temperature (heat transfer) in a dynamic fluid flow.

Completing the Taylor Series expansion for the Lax Finite Difference scheme and determining the consistency, enables the Truncation Error and order of accuracy to be determined.

4.4.2 Truncation Error and Order of Accuracy of the Lax Scheme

The completed derivation of Taylor Series expansion of the Lax equation enables the Truncation Error of the scheme to be quantified by the following Truncation Error terms:

$$T.E. = \frac{\Delta t}{2!}\left.\frac{\partial^2 T}{\partial t^2}\right|_i^n + \frac{\Delta t^2}{3!}\left.\frac{\partial^3 T}{\partial t^3}\right|_i^n - \frac{\Delta x^2}{2\Delta t}\left.\frac{\partial^2 T}{\partial x^2}\right|_i^n + U\frac{\Delta x^2}{3!}\left.\frac{\partial^3 T}{\partial x^3}\right|_i^n \qquad (4.15)$$

Hence, the leading Truncation Error terms in time and space for the Lax scheme are of order Δt and Δx^2 respectively. As a result, the Lax Scheme is theoretically first order accurate in time and second order accurate in space. However, the Lax scheme has been proved to be **inconsistent** so it will never be convergent. Consequently, the accuracy of the numerical solutions will never be 100% valid.

4.5 Taylor Series Expansion of the Lax-Wendroff Scheme

To enable the Taylor Series expansion for Lax-Wendroff scheme to be derived it is necessary to rearrange the initial discretised equation (equation 1.3) so that it is equal to zero.

$$\left(\frac{T_i^{n+1} - T_i^n}{\Delta t}\right) + U\left(\frac{T_{i+1}^n - T_{i-1}^n}{2\Delta x}\right) - U^2 \frac{\Delta t}{2}\left(\frac{T_{i+1}^n - 2T_i^n + T_{i-1}^n}{\Delta x^2}\right)$$
$$= 0 \qquad (4.16)$$

Consequently the Lax-Wendroff Taylor Series expansion can be derived. The initial Taylor Series expansion for the Lax-Wendroff is given by the following:

$$\frac{1}{\Delta t}\left(T_i^n + \Delta t \left.\frac{\partial T}{\partial t}\right|_i^n + \frac{\Delta t^2}{2!} \left.\frac{\partial^2 T}{\partial t^2}\right|_i^n + \frac{\Delta t^3}{3!} \left.\frac{\partial^3 T}{\partial t^3}\right|_i^n + \cdots - T_i^n\right)$$

$$+ \frac{U}{2\Delta x}\left(\left(T_i^n + \Delta x \left.\frac{\partial T}{\partial x}\right|_i^n + \frac{\Delta x^2}{2!} \left.\frac{\partial^2 T}{\partial x^2}\right|_i^n + \frac{\Delta x^3}{3!} \left.\frac{\partial^3 T}{\partial x^3}\right|_i^n + \cdots\right)\right.$$

$$\left. - \left(T_i^n - \Delta x \left.\frac{\partial T}{\partial x}\right|_i^n + \frac{\Delta x^2}{2!} \left.\frac{\partial^2 T}{\partial x^2}\right|_i^n - \frac{\Delta x^3}{3!} \left.\frac{\partial^3 T}{\partial x^3}\right|_i^n + \cdots\right)\right)$$

$$- U^2 \frac{\Delta t}{2\Delta x^2}\left(\left(T_i^n + \Delta x \left.\frac{\partial T}{\partial x}\right|_i^n + \frac{\Delta x^2}{2!} \left.\frac{\partial^2 T}{\partial x^2}\right|_i^n + \frac{\Delta x^3}{3!} \left.\frac{\partial^3 T}{\partial x^3}\right|_i^n + \cdots\right)\right.$$

$$\left. - 2T_i^n + \left(T_i^n - \Delta x \left.\frac{\partial T}{\partial x}\right|_i^n + \frac{\Delta x^2}{2!} \left.\frac{\partial^2 T}{\partial x^2}\right|_i^n - \frac{\Delta x^3}{3!} \left.\frac{\partial^3 T}{\partial x^3}\right|_i^n + \cdots\right)\right)$$

$$= 0 \qquad (4.17)$$

Rearranging and simplifying this initial Taylor Series expansion (equation 4.17) enables a streamlined version of the Lax-Wendroff Taylor Series expansion equation to be obtained.

The rearranged expansion is;

$$\left(\frac{\partial T}{\partial t}\bigg|_i^n + \frac{\Delta t}{2!}\frac{\partial^2 T}{\partial t^2}\bigg|_i^n + \frac{\Delta t^2}{3!}\frac{\partial^3 T}{\partial t^3}\bigg|_i^n\right)$$

$$+ \frac{U}{2\Delta x}\left(2\Delta x \frac{\partial T}{\partial x}\bigg|_i^n + \frac{2\Delta x^3}{3!}\frac{\partial^3 T}{\partial x^3}\bigg|_i^n\right) - U^2 \frac{\Delta t}{2\Delta x^2}\left(\frac{2\Delta x^2}{2!}\frac{\partial^2 T}{\partial x^2}\bigg|_i^n\right)$$

$$= 0 \tag{4.18}$$

The final Taylor series expansion for the Lax-Wendroff Finite Difference equation is;

$$\frac{\partial T}{\partial t}\bigg|_i^n + U\frac{\partial T}{\partial x}\bigg|_i^n + \frac{\Delta t}{2!}\frac{\partial^2 T}{\partial t^2}\bigg|_i^n - U^2\frac{\Delta t}{2!}\frac{\partial^2 T}{\partial x^2}\bigg|_i^n + \frac{\Delta t^2}{3!}\frac{\partial^3 T}{\partial t^3}\bigg|_i^n + \frac{\Delta x^2}{3!}\frac{\partial^3 T}{\partial x^3}\bigg|_i^n$$

$$= 0 \tag{4.19}$$

Deriving the full Taylor Series expansion of the Lax-Wendroff Finite Difference scheme enables the consistency of the scheme to be investigated.

4.5.1 Consistency Verification of the Lax-Wendroff Scheme

As Δt and Δx tend to zero the Taylor Series expansion for the Lax-Wendroff equation reduces to the following differential equation evaluated in terms of points i, n:

$$\frac{\partial T}{\partial t}\bigg|_i^n + U\frac{\partial T}{\partial x}\bigg|_i^n = 0 \tag{4.20}$$

It can be observed that as Δx and Δt tend to zero the Lax-Wendroff Taylor Series expansion returns to the original partial differential equation (equation 2.0). Therefore, the Lax-Wendroff Finite Difference scheme is **Consistent**.

Completing the Taylor Series expansion for the Lax Wendroff scheme and determining the consistency, enables the Truncation Error and order of accuracy of the Finite Difference scheme to be determined.

4.5.2 Truncation Error and Order of Accuracy of the Lax-Wendroff Scheme

The Taylor Series expansion of the Lax-Wendroff equation enables the Truncation Error of this scheme to be quantified by the flowing Truncation Error terms:

$$T.E. = \frac{\Delta t}{2!}\frac{\partial^2 T}{\partial t^2}\bigg|_i^n - U^2 \frac{\Delta t}{2!}\frac{\partial^2 T}{\partial x^2}\bigg|_i^n + \frac{\Delta t^2}{3!}\frac{\partial^3 T}{\partial t^3}\bigg|_i^n + \frac{\Delta x^2}{3!}\frac{\partial^3 T}{\partial x^3}\bigg|_i^n \quad (4.21)$$

Equation 4.21 shows that the leading Truncation Error terms in time and space for the Lax-Wendroff scheme are of order Δt and Δx^2, respectively. Therefore, it is first order accurate in time and second order accurate in space.

Completion of the derivations of Taylor Series expansions, Consistency Verifications, Truncation Error quantification and, determining the Order of Accuracy for the FTBS, Lax and Lax-Wendroff Finite Difference schemes, enables the numerical solutions for the original partial differential equation (equation 2.0) for the linear convection of temperature in a fluid flow, to be calculated using the Computational Domain Grid System for the three schemes.

Numerical Solutions and Performance Analysis

This chapter details the numerical solutions for the linear convection of temperature (heat transfer) for the FTBS, Lax and Lax-Wendroff schemes. Furthermore, it also showcases the graphical plots for each subsequent time-step point downstream in the fluid flow, showing the progression of heat downstream as time goes on.

A comprehensive performance analysis of each of the three Finite Difference schemes is conducted based on the numerical solutions, in conjunction with the results of the Taylor Series expansions consistency verifications. Thus, fulfilling the investigation objectives. Consequently, enabling a thorough discussion and conclusions about the usefulness of the Finite Difference method for solving partial differential equations of linear processes to be determined.

5.1 Forward-Time, Backwards-Space Scheme

This subsection details the numerical solutions, graphical outputs and, the performance analysis of the Forward-Time, Backwards-Space Finite Difference scheme.

5.1.1 Numerical Solutions and Graphical Outputs

The following showcases the numerical solutions and graphical outputs from the FTBS scheme that were obtained for the linear convection of temperature partial differential equation as part of the performance analysis investigation.

5.1.1.1 Numerical Solutions for a Time-Step Size of 0.01s

Table 5.0 shows the Numerical Solutions for the FTBS scheme, which were calculated using the Finite Difference Computational Grid System (figure 3.0), for a time-step size of 0.01s ($\Delta t = \mathbf{0.01}$ and consequently $v = \mathbf{0.05}$).

x-direction		n	n+1	n+2	n+3	n+4	n+5	n+6	n+7	n+8	n+9	n+10	n+11
0	i	0	0	0	0	0	0	0	0	0	0	0	0
0.1	i+1	1	0.9	0.81	0.729	0.6561	0.59049	0.531441	0.478297	0.430467	0.38742	0.348678	0.313811
0.2	i+2	2	1.9	1.8	1.701	1.6038	1.50903	1.417176	1.328603	1.243572	1.162261	1.084777	1.011167
0.3	i+3	2	2	1.99	1.971	1.944	1.90998	1.869885	1.824614	1.775013	1.721869	1.665908	1.607795
0.4	i+4	1	1.1	1.19	1.27	1.3401	1.40049	1.451439	1.493284	1.526417	1.551276	1.568336	1.578093
0.5	i+5	0	0.1	0.2	0.299	0.3961	0.4905	0.581499	0.668493	0.750972	0.828517	0.900792	0.967547
0.6	i+6	0	0	0.01	0.029	0.056	0.09001	0.130059	0.175203	0.224532	0.277176	0.33231	0.389158
0.7	i+7	0	0	0	0.001	0.0038	0.00902	0.017119	0.028413	0.043092	0.061236	0.08283	0.107778
0.8	i+8	0	0	0	0	0.0001	0.00047	0.001325	0.002904	0.005455	0.009219	0.014421	0.021262
0.9	i+9	0	0	0	0	0	0.00001	0.000056	0.000183	0.000455	0.000955	0.001781	0.003045
1	i+10	0	0	0	0	0	0	0.000001	6.5E-06	2.41E-05	6.72E-05	0.000156	0.000319
1.1	i+11	0	0	0	0	0	0	0	1E-07	7.4E-07	3.08E-06	9.5E-06	2.41E-05
1.2	i+12	0	0	0	0	0	0	0	0	1E-08	8.3E-08	3.83E-07	1.29E-06
1.3	i+13	0	0	0	0	0	0	0	0	0	1E-09	9.2E-09	4.66E-08
1.4	i+14	0	0	0	0	0	0	0	0	0	0	1E-10	1.01E-09
1.5	i+15	0	0	0	0	0	0	0	0	0	0	0	1E-11
1.6	i+16	0	0	0	0	0	0	0	0	0	0	0	0
1.7	i+17	0	0	0	0	0	0	0	0	0	0	0	0
1.8	i+18	0	0	0	0	0	0	0	0	0	0	0	0
1.9	i+19	0	0	0	0	0	0	0	0	0	0	0	0
2	i+20	0	0	0	0	0	0	0	0	0	0	0	0
2.1	i+21	0	0	0	0	0	0	0	0	0	0	0	0

Table 5.0: The Numerical Solutions for the FTBS Finite Difference Scheme when $\Delta t = 0.01$ and $v = 0.05$.

Chapter 5   Performance Analysis

5.1.1.2 Graphical Outputs for a Time-Step Size of 0.01s

Figure 5.0 shows (as a series of graphical outputs) the temperature convection downstream in a fluid flow as time progresses in intervals 0.01s, for the FTBS scheme, for a time-step size of 0.01s (Δt = **0.01** and consequently v = **0.05**).

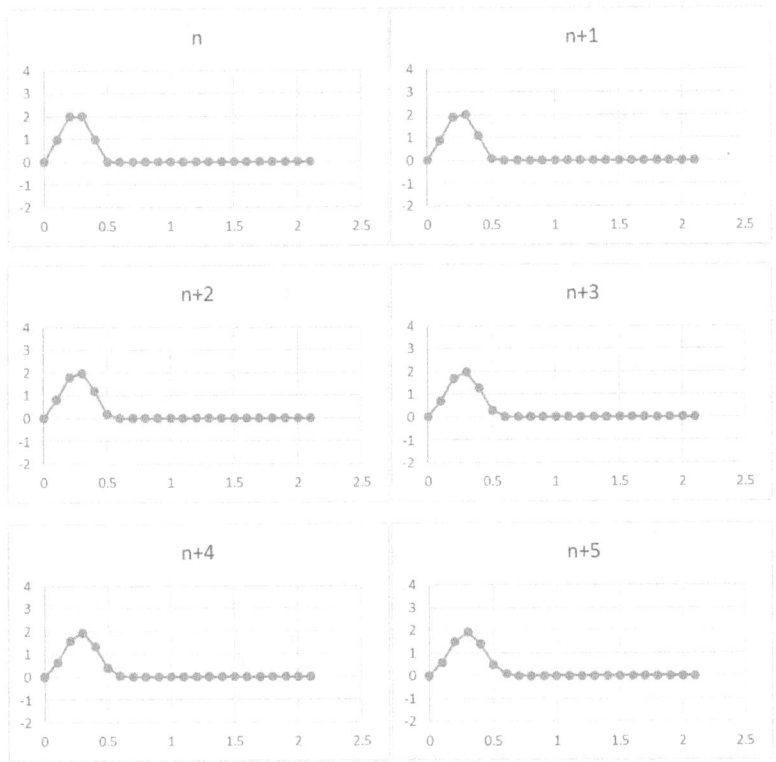

Chapter 5	Performance Analysis

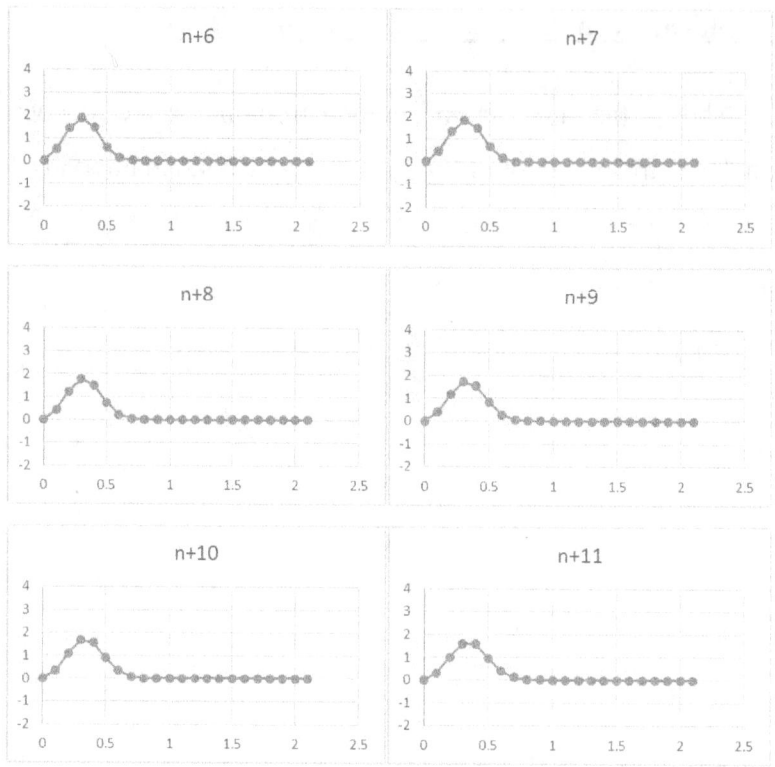

Figure 5.0: Graphs showing the Convection of Temperature Downstream in a Fluid Flow for the FTBS Finite Difference Scheme at the 12 Computational Grid Points (Time Intervals of 0.01s) when $\Delta t = 0.01$ and $v = 0.05$.

Chapter 5 — Performance Analysis

5.1.1.3 Numerical Solutions for a Time-Step Size of 0.1s

Table 5.1 shows the Numerical Solutions for the FTBS scheme, which were calculated using the Finite Difference Computational Grid System (figure 3.0), for a time-step size of 0.1s ($\Delta t = 0.1$ and consequently $v = 0.05$).

x-direction		n	n+1	n+2	n+3	n+4	n+5	n+6	n+7	n+8	n+9	n+10	n+11
0	i	0	0	0	0	0	0	0	0	0	0	0	0
0.1	i+1	1	0	0	0	0	0	0	0	0	0	0	0
0.2	i+2	2	1	0	0	0	0	0	0	0	0	0	0
0.3	i+3	2	2	1	0	0	0	0	0	0	0	0	0
0.4	i+4	1	2	2	1	0	0	0	0	0	0	0	0
0.5	i+5	0	1	2	2	1	0	0	0	0	0	0	0
0.6	i+6	0	0	1	2	2	1	0	0	0	0	0	0
0.7	i+7	0	0	0	1	2	2	1	0	0	0	0	0
0.8	i+8	0	0	0	0	1	2	2	1	0	0	0	0
0.9	i+9	0	0	0	0	0	1	2	2	1	0	0	0
1	i+10	0	0	0	0	0	0	1	2	2	1	0	0
1.1	i+11	0	0	0	0	0	0	0	1	2	2	1	0
1.2	i+12	0	0	0	0	0	0	0	0	1	2	2	1
1.3	i+13	0	0	0	0	0	0	0	0	0	1	2	2
1.4	i+14	0	0	0	0	0	0	0	0	0	0	1	2
1.5	i+15	0	0	0	0	0	0	0	0	0	0	0	1
1.6	i+16	0	0	0	0	0	0	0	0	0	0	0	0
1.7	i+17	0	0	0	0	0	0	0	0	0	0	0	0
1.8	i+18	0	0	0	0	0	0	0	0	0	0	0	0
1.9	i+19	0	0	0	0	0	0	0	0	0	0	0	0
2	i+20	0	0	0	0	0	0	0	0	0	0	0	0
2.1	i+21	0	0	0	0	0	0	0	0	0	0	0	0

Table 5.1: The Numerical Solutions for the FTBS Finite Difference Scheme when $\Delta t = 0.1$ and $v = 0.5$.

5.1.1.4 Graphical Outputs for a Time-Step Size of 0.1s

Figure 5.1 shows (as a series of graphical outputs) the temperature convection downstream in a fluid flow as time progresses in intervals 0.1s, for the FTBS scheme, for a time-step size of 0.1s ($\Delta t = 0.1$ and consequently $v = 0.5$).

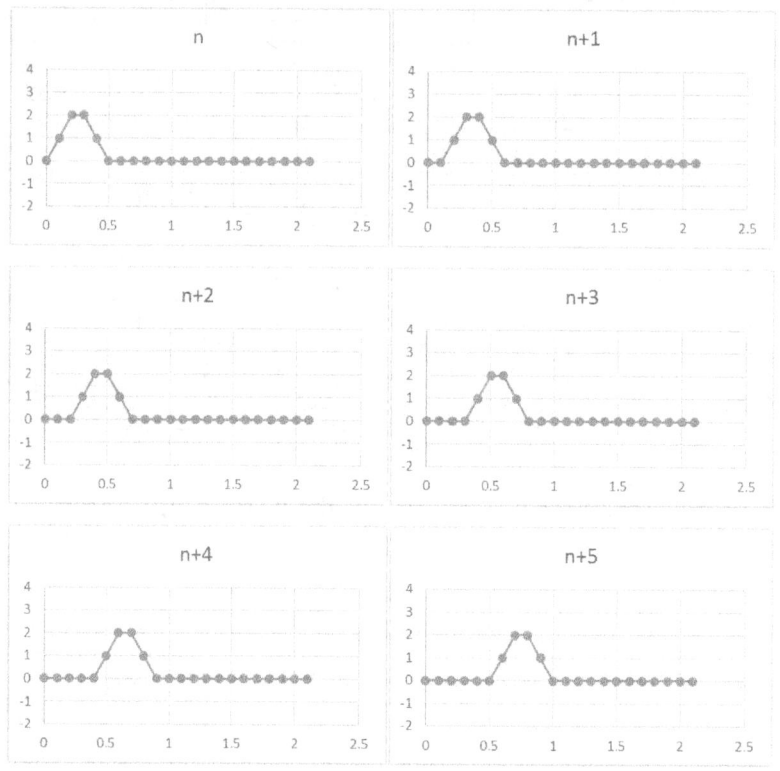

Chapter 5   Performance Analysis

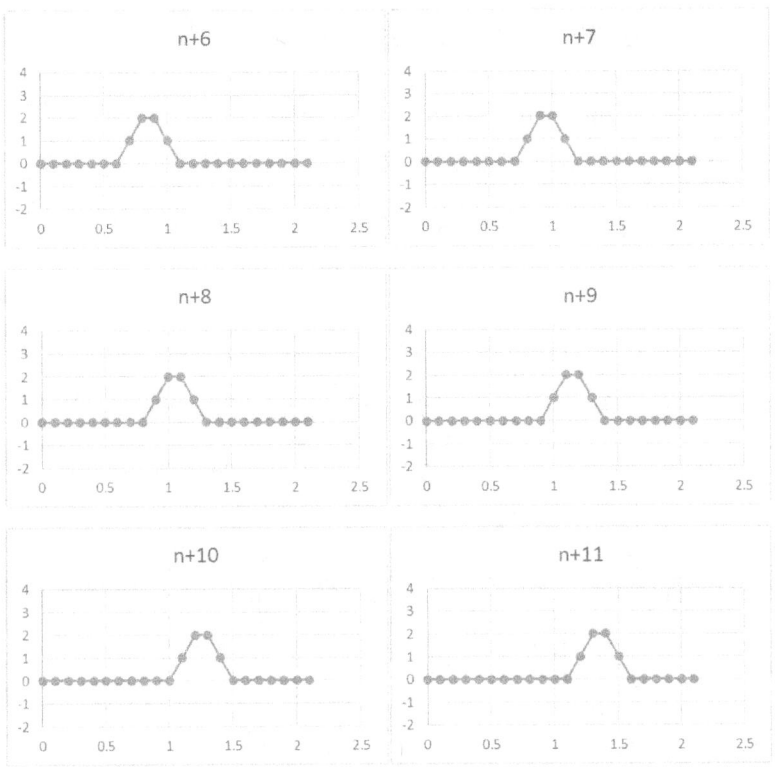

Figure 5.1: Graphs showing the Convection of Temperature Downstream in a Fluid Flow for the FTBS Finite Difference Scheme at the 12 Computational Grid Points (Time Intervals of 0.1s) when $\Delta t = 0.1$ and $v = 0.5$.

5.1.1.5 Numerical Solutions for a Time-Step Size of 0.12s

Table 5.2 shows the Numerical Solutions for the FTBS scheme, which were calculated using the Finite Difference Computational Grid System (figure 3.0), for a time-step size of 0.12s ($\Delta t = 0.12$ and consequently $v = 0.6$).

x-direction	i	n	n+1	n+2	n+3	n+4	n+5	n+6	n+7	n+8	n+9	n+10	n+11
0	i	0	0	0	0	0	0	0	0	0	0	0	0
0.1	i+1	1	-0.2	0.04	-0.008	0.0016	-0.00032	6.4E-05	-1.3E-05	2.56E-06	-5.1E-07	1.02E-07	-2E-08
0.2	i+2	2	0.8	-0.4	0.128	-0.0352	0.00896	-0.00218	0.000512	-0.00012	2.66E-05	-5.9E-06	1.31E-06
0.3	i+3	2	2	0.56	-0.592	0.272	-0.09664	0.03008	-0.00863	0.00234	-0.00061	0.000154	-3.8E-05
0.4	i+4	1	2.2	1.96	0.28	-0.7664	0.47968	-0.2119	0.078477	-0.02605	0.008017	-0.00233	0.000651
0.5	i+5	0	1.2	2.4	1.872	-0.0384	-0.912	0.758016	-0.40589	0.17535	-0.06633	0.022886	-0.00738
0.6	i+6	0	0	1.44	2.592	1.728	-0.39168	-1.01606	1.112832	-0.70963	0.352346	-0.15006	0.057476
0.7	i+7	0	0	0	1.728	2.7648	1.52064	-0.77414	-1.06445	1.548288	-1.16122	0.655059	-0.31109
0.8	i+8	0	0	0	0	2.0736	2.90304	1.24416	-1.1778	-1.04178	2.066301	-1.80672	1.147414
0.9	i+9	0	0	0	0	0	2.48832	2.985984	0.895795	-1.59252	-0.93163	2.665887	-2.70124
1	i+10	0	0	0	0	0	0	2.985984	2.985984	0.477757	-2.00658	-0.71664	3.342391
1.1	i+11	0	0	0	0	0	0	0	3.583181	2.866545	0	-2.4079	-0.37838
1.2	i+12	0	0	0	0	0	0	0	0	4.299817	2.57989	-0.51598	-2.78628
1.3	i+13	0	0	0	0	0	0	0	0	0	5.15978	2.063912	-1.03196
1.4	i+14	0	0	0	0	0	0	0	0	0	0	6.191736	1.238347
1.5	i+15	0	0	0	0	0	0	0	0	0	0	0	7.430084
1.6	i+16	0	0	0	0	0	0	0	0	0	0	0	0
1.7	i+17	0	0	0	0	0	0	0	0	0	0	0	0
1.8	i+18	0	0	0	0	0	0	0	0	0	0	0	0
1.9	i+19	0	0	0	0	0	0	0	0	0	0	0	0
2	i+20	0	0	0	0	0	0	0	0	0	0	0	0
2.1	i+21	0	0	0	0	0	0	0	0	0	0	0	0

Table 5.2: The Numerical Solutions for the FTBS Finite Difference Scheme when $\Delta t = 0.12$ and $v = 0.6$.

Chapter 5   Performance Analysis

5.1.1.6 Graphical Outputs for a Time-Step Size of 0.12s

Figure 5.2 shows (as a series of graphical outputs) the temperature convection downstream in a fluid flow as time progresses in intervals 0.12s, for the FTBS scheme, for a time-step size of 0.12s ($\Delta t = 0.12$ and consequently $v = 0.6$).

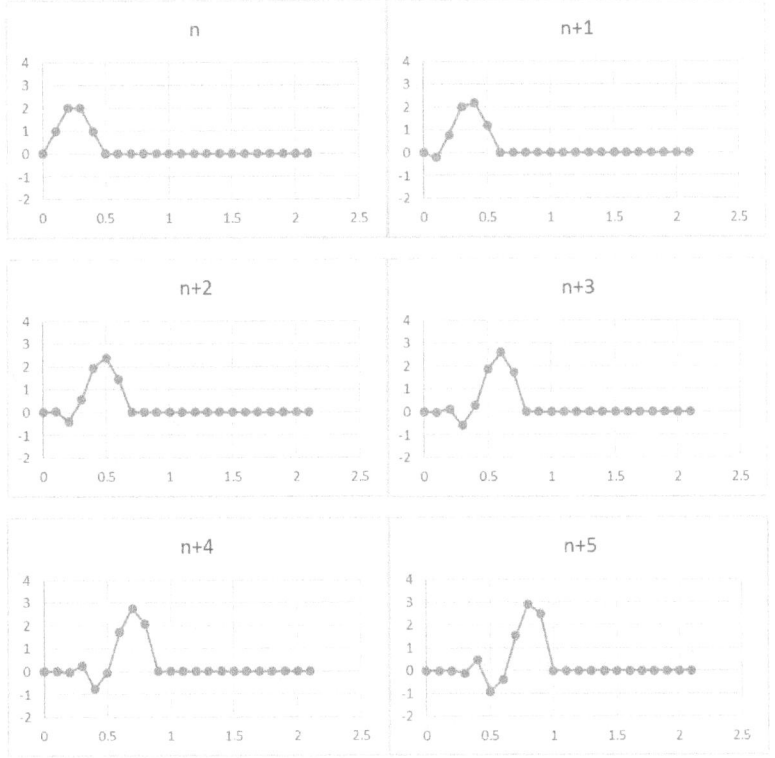

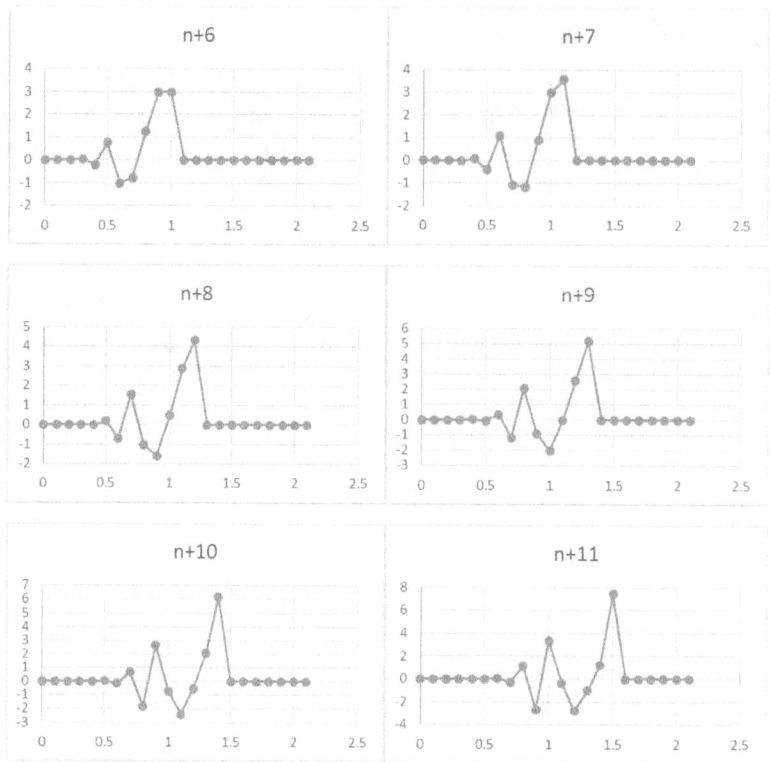

Figure 5.2: Graphs showing the Convection of Temperature Downstream in a Fluid Flow for the FTBS Finite Difference Scheme at the 12 Computational Grid Points (Time Intervals of 0.12s) when $\Delta t = 0.12$ and $v = 0.6$.

5.1.2 FTBS Scheme Performance Analysis

It can be observed from the FTBS scheme graphical results (figure 5.0) for a time step of 0.01s (and v equal to 0.05), that this Finite Difference scheme produces stable results for the linear convection of temperature in a fluid flow traveling with a constant convection velocity. This time step size causes the numerical solutions to be inaccurate for the temperature distribution across the fluid flow, because there is minimal deviation from the initial conditions across the time intervals. Although the results are stable for a time step size of 0.01s, it can be observed that the temperature distribution remains relatively unchanged across all time intervals since there is very minimal progression of the peak temperature in the x-direction.

Furthermore, the graphical outputs highlight that downstream convection of the temperature has begun to commence, because the peak temperature value has proceeded to diminish slightly throughout the twelve time intervals. Thus, reducing from the peak temperature at the initial conditions, which was value of two. Hence, it can be determined that time step size of 0.01s is excessively small for this type of Finite Difference scheme, which derives solutions using a backwards difference technique. This is verified as the change in temperature distribution in the x-direction is negligible, given that the numerical outputs struggle to show the convection of temperature in the fluid flow. Occurring due to the overall length of time that is assessed being only 0.12s, which is effectively only covering the start of the heat transfer downstream of the fluid flow.

Implementing such a small-time step has rendered the FTBS scheme's ability to provide an accurate representation of the linear convection of temperature obsolete across the full fluid flow in the x-direction. Therefore, this Finite Difference scheme is unable to provide suitable numerical results when a very small-time step of the magnitude of 1/100th of a second is applied. Consequently, it is advisable not to operate the FTBS scheme with a time-step size of the order of 0.01s.

Having determined that the FTBS scheme cannot operate effectively with a very minuscule time step, the application of a larger time step size is required to improve performance. A time step of 0.1s was subsequently implemented into the FTBS Finite Difference Computational grid system. It should be noted that this time-step size is the limit of stability for not only the FTBS scheme, it is the limit of stability for all the three Finite Difference schemes. Because it accurately shows the temperature profile moving linearly downstream as shown in figure 5.1. Therefore, a time-step of 0.1s provides accurate and stable numerical solutions for the temperature distribution in the fluid flow as time progresses. This correlates with the linear relationship of the initial partial differential equation. Hence, it was expected that the heat transfer would move uniformly in the fluid flow along the x-direction and throughout the computational grid system.

It can be observed from the graphical outputs of the FTBS scheme (figure 5.1) when Δt is equal to 0.1 that the temperature distribution in the fluid flow progresses linearly in the x-direction downstream of the fluid flow. At each of the twelve time intervals there is no change to the numerical value of the temperature at the initial conditions. Furthermore, figure 5.1 highlights that the constant convection velocity of the system enables the temperature distribution to remain consistent as it travels downstream of the fluid. Therefore, conserving the linear relationship of the original partial differential equation.

The numerical solutions show that the FTBS scheme is stable when the time step is 0.1s and v equals 0.5. Given that there is no fluctuations in the temperature distribution whilst the peak temperature remains constant as the temperature moves downstream throughout the fluid. Hence, the numerical solutions obtained from the Finite Difference scheme are accurate. Thus, when Δt equals 0.1 (v equals 0.5) the FTBS Finite Difference scheme is accurate and stable. Since the graphical representation of the numerical solution outputs evidently highlight linear convection of temperature through the fluid flow in the x-direction as time progresses.

The final value of v used in the investigation of the performance of the FTBS scheme was 0.6 which equates to a time-step size of 0.12s. This time-step size produces very interesting results for the performance of the FTBS Finite Difference scheme when applied to the partial differential equation that describes the linear convection of temperature in a fluid flow. Observations of the graphical output results for this time step size (figure 5.2) highlight that the scheme starts off relatively stable with only the presence of minor negative values of temperature at the time intervals n+1 to n+3.

However, as the temperature distribution in the fluid flow progresses downstream, increasing amounts of distortions and fluctuations occur with increasing amplitude, causing significant instability to be present. Consequently, the FTBS Finite Difference scheme and the subsequent numerical solutions become considerably unstable as the temperature undergoes convection through the fluid flow, especially at later time intervals.

Additionally, the peak temperature value (when Δt = 0.12s) continually increases as the temperature distribution moves downstream in the fluid to a maximum value of 7.43 at the final time interval n+11. Indicating that the FTBS scheme is inaccurate as well as unstable for this magnitude of time-step. Showing that the effects of truncation and round off errors in the numerical solutions are increasing as the scheme progresses. As a result, the instability of the Finite Difference scheme causes continual increases to the amounts of fluctuations in the solutions. Further exacerbating the inaccuracy of the numerical solutions and the scheme as time progresses.

Moreover, the presence of negative temperatures in the numerical solutions highlight that the FTBS scheme is unable to accurately represent nor model the process of linear convection successfully, when Δt equals 0.6. Diminishing its ability to accurately provide numerical solutions for the original partial differential equation when the time-step is equal to 0.12s. Due to instability. As a result, the FTBS scheme suffers a considerable depreciation to its performance.

Overall the graphical outputs (figure 5.2) highlight that for a time-step size of 0.12s the FTBS Finite Difference scheme is unstable. Thus, providing inaccurate numerical solutions for the linear convection of temperature in a fluid flow. Furthermore, the instability in the results continually grows, which has a more profound effect on the accuracy of the temperature solutions as the fluid progress downstream from its initial conditions.

Consequently, for optimal performance the FTBS Finite Difference scheme, should operate with a time step size of 0.1s when employed to provide numerical solutions for linear partial differential equations.

Chapter 5　　　　　　　　　　　　　　　　　　　　　　　Performance Analysis

5.2 Lax Scheme

This subsection details the numerical solutions, graphical outputs and, the performance analysis of the Lax Finite Difference scheme.

5.2.1 Numerical Solutions and Graphical Outputs

The following highlights the numerical solutions and graphical outputs from the Lax scheme that were obtained for the linear convection of temperature partial differential equation as part of the performance analysis investigation.

5.2.1.1 Numerical Solutions for a Time-Step Size of 0.01s

Table 5.3 shows the Numerical Solutions for the Lax scheme, which were calculated using the Finite Difference Computational Grid System (figure 3.0), for a time-step size of 0.01s ($\Delta t = 0.01$ and consequently $v = 0.05$).

x-direction	i	n	n+1	n+2	n+3	n+4	n+5	n+6	n+7	n+8	n+9	n+10	n+11
0	i	0	0	0	0	0	0	0	0	0	0	0	0
0.1	i+1	1	0.9	0.6525	0.536625	0.423225	0.365867	0.299084	0.269175	0.224457	0.208152	0.175778	0.166748
0.2	i+2	2	1.45	1.1925	0.9405	0.813038	0.66463	0.598167	0.498794	0.46256	0.390617	0.370552	0.315666
0.3	i+3	2	1.55	1.2925	1.150875	0.959681	0.88209	0.742885	0.698918	0.593702	0.569041	0.486639	0.473498
0.4	i+4	1	1.1	1.1	0.983125	0.966488	0.83853	0.822059	0.7097	0.699184	0.604	0.599322	0.518788
0.5	i+5	0	0.55	0.605	0.741125	0.690456	0.748688	0.669139	0.699508	0.616586	0.636332	0.558081	0.573726
0.6	i+6	0	0	0.3025	0.33275	0.482488	0.462107	0.549724	0.502781	0.559514	0.501957	0.542442	0.482232
0.7	i+7	0	0	0	0.166375	0.183013	0.306546	0.299454	0.38841	0.361855	0.427687	0.389528	0.440404
0.8	i+8	0	0	0	0	0.091506	0.100657	0.191248	0.189612	0.266565	0.252115	0.315692	0.291306
0.9	i+9	0	0	0	0	0	0.050328	0.055361	0.117643	0.117989	0.17881	0.171257	0.226619
1	i+10	0	0	0	0	0	0	0.027681	0.030449	0.071554	0.07243	0.117751	0.113983
1.1	i+11	0	0	0	0	0	0	0	0.015224	0.016747	0.043123	0.043981	0.076368
1.2	i+12	0	0	0	0	0	0	0	0	0.008373	0.009211	0.02579	0.026469
1.3	i+13	0	0	0	0	0	0	0	0	0	0.004605	0.005066	0.015324
1.4	i+14	0	0	0	0	0	0	0	0	0	0	0.002533	0.002786
1.5	i+15	0	0	0	0	0	0	0	0	0	0	0	0.001393
1.6	i+16	0	0	0	0	0	0	0	0	0	0	0	0
1.7	i+17	0	0	0	0	0	0	0	0	0	0	0	0
1.8	i+18	0	0	0	0	0	0	0	0	0	0	0	0
1.9	i+19	0	0	0	0	0	0	0	0	0	0	0	0
2	i+20	0	0	0	0	0	0	0	0	0	0	0	0
2.1	i+21	0	0	0	0	0	0	0	0	0	0	0	0

Table 5.3: The Numerical Solutions for the Lax Finite Difference Scheme when $\Delta t = 0.01$ and $v = 0.05$.

Chapter 5   Performance Analysis

5.2.1.2 Graphical Outputs for a Time-Step Size of 0.01s

Figure 5.3 shows (as a series of graphical outputs) the temperature convection downstream in a fluid flow as time progresses in intervals 0.01s, for the Lax scheme, for a time-step size of 0.01s ($\Delta t = 0.01$ and consequently $v = 0.05$).

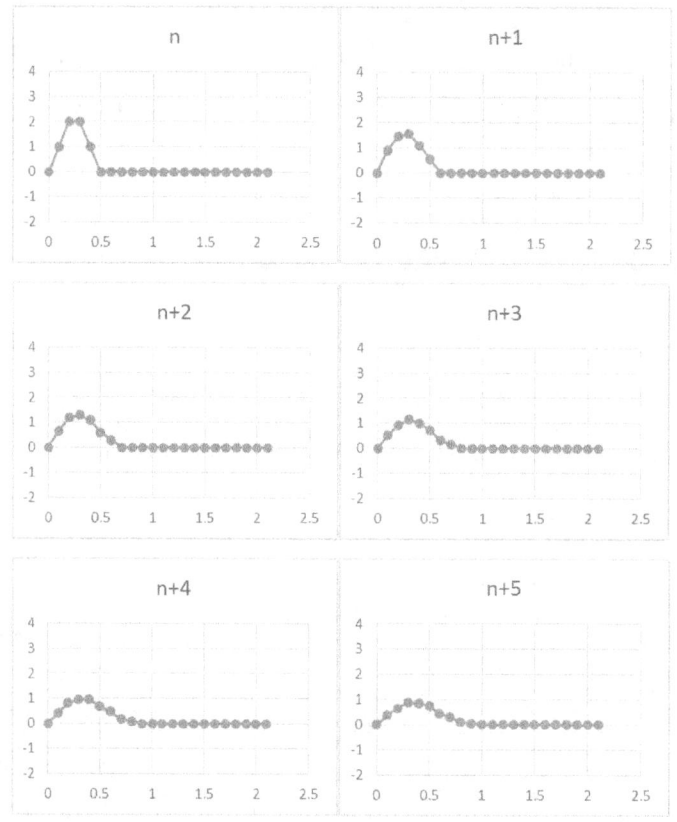

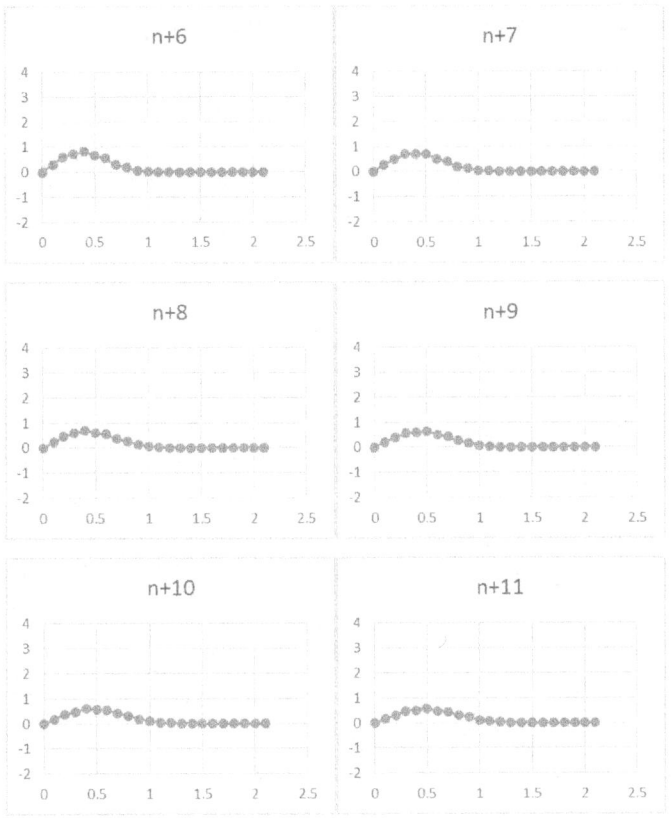

Figure 5.3: Graphs showing the Convection of Temperature Downstream in a Fluid Flow for the Lax Finite Difference Scheme at the 12 Computational Grid Points (Time Intervals of 0.01s) when $\Delta t = 0.01$ and $v = 0.05$.

5.2.1.3 Numerical Solutions for a Time-Step Size of 0.1s

Table 5.4 shows the Numerical Solutions for the Lax scheme, which were calculated using the Finite Difference Computational Grid System (figure 3.0), for a time-step size of 0.1s ($\Delta t = 0.1$ and consequently $v = 0.5$).

x-direction		n	n+1	n+2	n+3	n+4	n+5	n+6	n+7	n+8	n+9	n+10	n+11
0	i	0	0	0	0	0	0	0	0	0	0	0	0
0.1	i+1	1	0	0	0	0	0	0	0	0	0	0	0
0.2	i+2	2	1	0	0	0	0	0	0	0	0	0	0
0.3	i+3	2	2	1	0	0	0	0	0	0	0	0	0
0.4	i+4	1	2	2	1	0	0	0	0	0	0	0	0
0.5	i+5	0	1	2	2	1	0	0	0	0	0	0	0
0.6	i+6	0	0	1	2	2	1	0	0	0	0	0	0
0.7	i+7	0	0	0	1	2	2	1	0	0	0	0	0
0.8	i+8	0	0	0	0	1	2	2	1	0	0	0	0
0.9	i+9	0	0	0	0	0	1	2	2	1	0	0	0
1	i+10	0	0	0	0	0	0	1	2	2	1	0	0
1.1	i+11	0	0	0	0	0	0	0	1	2	2	1	0
1.2	i+12	0	0	0	0	0	0	0	0	1	2	2	1
1.3	i+13	0	0	0	0	0	0	0	0	0	1	2	2
1.4	i+14	0	0	0	0	0	0	0	0	0	0	1	2
1.5	i+15	0	0	0	0	0	0	0	0	0	0	0	1
1.6	i+16	0	0	0	0	0	0	0	0	0	0	0	0
1.7	i+17	0	0	0	0	0	0	0	0	0	0	0	0
1.8	i+18	0	0	0	0	0	0	0	0	0	0	0	0
1.9	i+19	0	0	0	0	0	0	0	0	0	0	0	0
2	i+20	0	0	0	0	0	0	0	0	0	0	0	0
2.1	i+21	0	0	0	0	0	0	0	0	0	0	0	0

Table 5.4: The Numerical Solutions for the Lax Finite Difference Scheme when $\Delta t = 0.1$ and $v = 0.5$.

Chapter 5   Performance Analysis

5.2.1.4 Graphical Outputs for a Time-Step Size of 0.1s

Figure 5.4 shows (as a series of graphical outputs) the temperature convection downstream in a fluid flow as time progresses in intervals 0.1s, for the Lax scheme, for a time-step size of 0.1s ($\Delta t = \mathbf{0.1}$ and consequently $v = \mathbf{0.5}$).

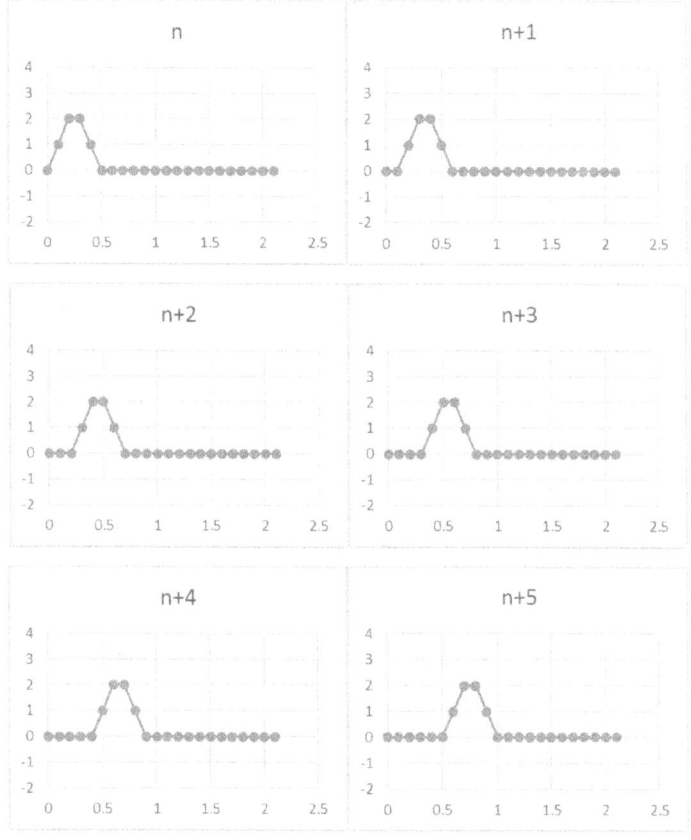

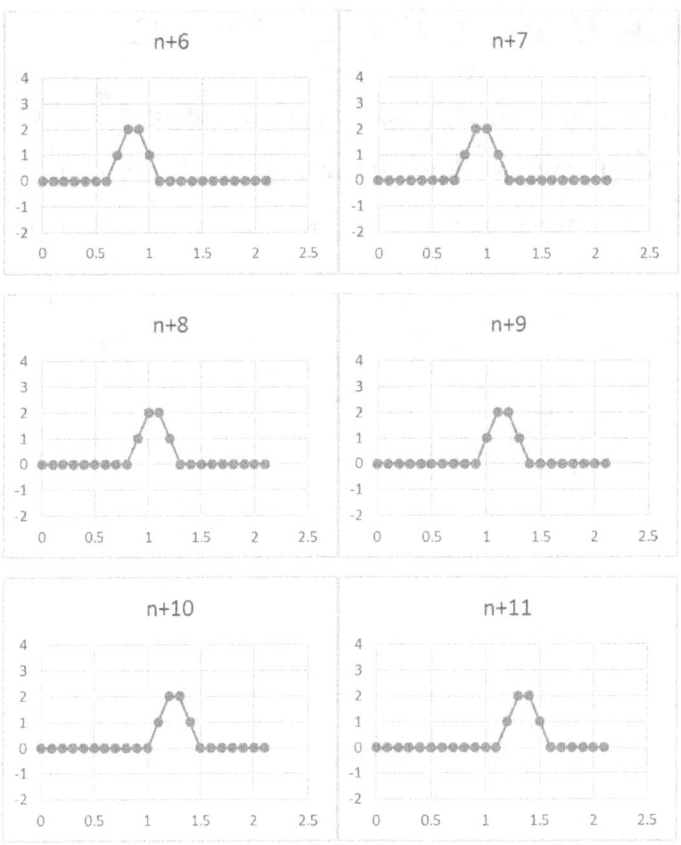

Figure 5.4: Graphs showing the Convection of Temperature Downstream in a Fluid Flow for the Lax Finite Difference Scheme at the 12 Computational Grid Points (Time Intervals of 0.1s) when $\Delta t = 0.1$ and $v = 0.5$.

Chapter 5	Performance Analysis

5.2.1.5 Numerical Solutions for a Time-Step Size of 0.12s

Table 5.5 shows the Numerical Solutions for the Lax scheme, which were calculated using the Finite Difference Computational Grid System (figure 3.0), for a time-step size of 0.12s ($\Delta t = 0.12$ and consequently $v = 0.6$).

x-direction		n	n+1	n+2	n+3	n+4	n+5	n+6	n+7	n+8	n+9	n+10	n+11
0	i	0	0	0	0	0	0	0	0	0	0	0	0
0.1	i+1	1	-0.2	-0.09	0.043	0.0176	-0.01166	-0.00448	0.003557	0.001304	-0.00117	-0.00041	0.000401
0.2	i+2	2	0.9	-0.43	-0.176	0.1166	0.04477	-0.03557	-0.01304	0.01166	0.004129	-0.00401	-0.00138
0.3	i+3	2	2.1	0.77	-0.693	-0.2541	0.22748	0.081191	-0.07746	-0.02694	0.027276	0.009277	-0.00987
0.4	i+4	1	2.2	2.2	0.605	-0.9922	-0.31944	0.383328	0.125913	-0.14451	-0.04735	0.054564	0.017733
0.5	i+5	0	1.1	2.42	2.299	0.3993	-1.331	-0.36603	0.59296	0.177156	-0.2456	-0.07529	0.09919
0.6	i+6	0	0	1.21	2.662	2.3958	0.14641	-1.713	-0.38652	0.866454	0.232074	-0.39169	-0.11147
0.7	i+7	0	0	0	1.331	2.9282	2.48897	-0.16105	-2.14198	-0.37203	1.215291	0.286461	-0.59611
0.8	i+8	0	0	0	0	1.4641	3.22102	2.576816	-0.53147	-2.62191	-0.31179	1.652512	0.3344
0.9	i+9	0	0	0	0	0	1.61051	3.543122	2.657342	-0.97436	-3.15692	-0.19292	2.192891
1	i+10	0	0	0	0	0	0	1.771561	3.897434	2.728204	-1.50051	-3.75128	0
1.1	i+11	0	0	0	0	0	0	0	1.948717	4.287178	2.786665	-2.12215	-4.40936
1.2	i+12	0	0	0	0	0	0	0	0	2.143589	4.715895	2.829537	-2.85312
1.3	i+13	0	0	0	0	0	0	0	0	0	2.357948	5.187485	2.853117
1.4	i+14	0	0	0	0	0	0	0	0	0	0	2.593742	5.706233
1.5	i+15	0	0	0	0	0	0	0	0	0	0	0	2.853117
1.6	i+16	0	0	0	0	0	0	0	0	0	0	0	0
1.7	i+17	0	0	0	0	0	0	0	0	0	0	0	0
1.8	i+18	0	0	0	0	0	0	0	0	0	0	0	0
1.9	i+19	0	0	0	0	0	0	0	0	0	0	0	0
2	i+20	0	0	0	0	0	0	0	0	0	0	0	0
2.1	i+21	0	0	0	0	0	0	0	0	0	0	0	0

Table 5.5: The Numerical Solutions for the Lax Finite Difference Scheme when $\Delta t = 0.12$ and $v = 0.6$.

5.2.1.6 Graphical Outputs for a Time-Step Size of 0.12s

Figure 5.5 shows (as a series of graphical outputs) the temperature convection downstream in a fluid flow as time progresses in intervals 0.12s, for the Lax scheme, for a time-step size of 0.12s ($\Delta t = \mathbf{0.12}$ and consequently $v = \mathbf{0.6}$).

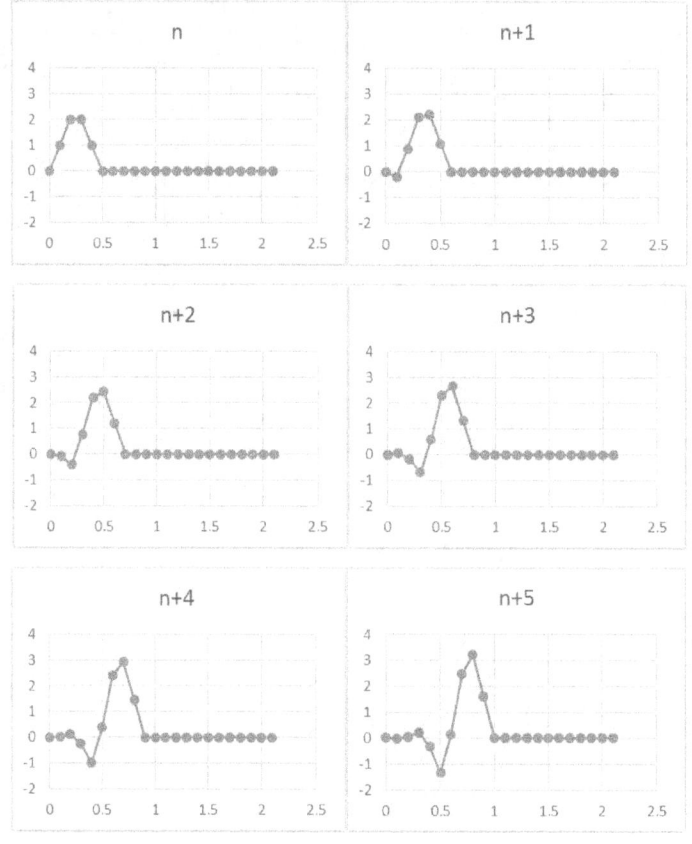

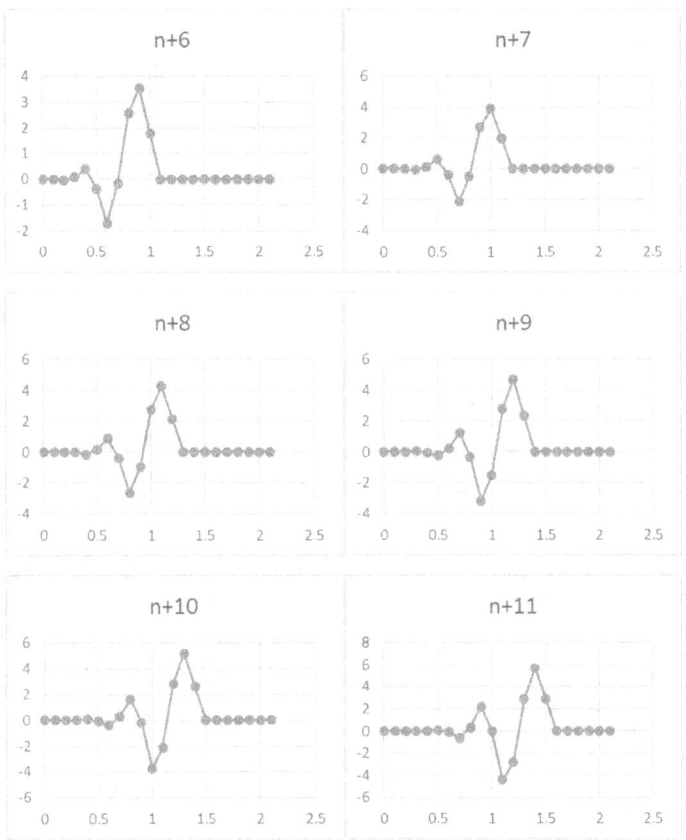

Figure 5.5: Graphs showing the Convection of Temperature Downstream in a Fluid Flow for the Lax Finite Difference Scheme at the 12 Computational Grid Points (Time Intervals of 0.12s) when $\Delta t = 0.12$ and $v = 0.6$.

5.2.2 Lax Scheme Performance Analysis

The results obtained from the investigation of the Lax Finite Difference scheme highlights several trends that can be deciphered through analysis.

Observations of the graphical outputs (figure 5.3) show that as the time intervals progress from n to n+11, the temperature distribution in the fluid does not move downstream along the x-direction from the initial conditions as expected. Instead the numerical solutions of the temperature just continuously reduce in magnitude from the peak value of two to below one. Suggesting that the temperature is dissipating in the fluid rather than travelling in a stable manner linearly downstream.

This is an inaccurate representation of the phenomena of linear convection, that is described by the original partial differential equation. Because, the temperature distribution should be consistently progressing downstream, as time moves forward. Consequently, showing that for a time step size of 0.01s the Lax Finite Difference scheme performs sub-optimally.

There are no fluctuations or visible numerical instability present in the numerical solutions (as shown by the graphical outputs (figure 5.3). Thus, it can be determined that the Lax scheme is stable when $v = 0.05$. Nonetheless, the scheme does not provide accurate consistent numerical solutions for this value of v. This is to be expected due to the presence of damping in the Lax scheme and small time-step size.

Moreover, the Lax Finite Difference scheme is inconsistent as it tends to zero, as proved by the Taylor Series expansion consistency verification. Therefore, a very small-time step will result in inconsistency of the results and a stunted progression of temperature downstream. Causing the temperature distribution to reduce from the peak at the initial conditions to nothing as shown in figure 5.3. Consequently, when the time-step size is 0.01s the Lax Finite Difference scheme is stable (as highlighted by the stable numerical solutions), but it is inaccurate.

The numerical solutions obtained from the Lax Finite Difference scheme for a time-step size of 0.1s, are identical to those obtained from the FTBS scheme when using the same time-step size. Thus, highlighting that the Lax scheme has an indistinguishable level of numerical performance from that observed from the FTBS scheme for this value of Δt. Hence, the graphical outputs (figure 5.4) clearly show the linear convection of the temperature through the fluid as time advances, because the peak temperature moves downstream of the fluid at each at each time interval.

Therefore, the Lax scheme performs optimally by providing an accurate representation of the model partial differential equation for the process of linear convection of temperature in a fluid flow. As a result the Lax Finite Difference scheme is stable and accurate when a time-step size of 0.1s is employed. Consequentially the numerical solutions have the same characteristics throughout the fluid. Thus, the continuity of the partial differential equation is preserved.

When Δt is equal to 0.12 the graphical outputs (figure 5.5) from the Lax scheme display numerous trends that can be analysed to determine the stability and accuracy of the numerical solutions. The temperature distribution shown by Figure 5.5 highlights considerable amounts of instability in the system and results. This is illustrated by the numerous fluctuations present in the graphical results. Figure 5.5 also shows that the instability of the Lax scheme is continuously increasing as the temperature profile moves downstream in the x-direction. Thus, intensifying the inaccuracy of the numerical solutions.

Additionally, the peak temperature is shown to be continually rising at each subsequent interval of time to a peak of 5.71, which when coupled with the presence of negative temperature results further highlights the inaccuracy of the scheme using a time-step size of 0.12s. Although the scheme is unstable, it is clear that the peak of the temperature distribution is progressing through the fluid linearly. Therefore, with a time step of 0.12s the Lax scheme still has a partial resemblance of ability to model the process of linear convection.

Overall, the Lax Finite Difference scheme is not only inconsistent as the time step size tends to zero, it similarly becomes progressively unstable and inaccurate for a time step size of 0.12s. As verified by thorough analysis of the graphical outputs results (figure 5.5). As a consequence, the scheme becomes more unstable and inaccurate if the time-step size in increased further. Hence, its performance will be significantly diminished as time step size increases.

Consequently, for optimised performance the Lax Finite Difference scheme should operate with a time-step size of 0.1s, when employed to model a process that is described by linear partial differential equations. Nevertheless, because the Lax scheme is inconsistent it cannot provide optimal performance nor achieve a wholly accurate representation of the linear process of heat moving through a fluid flow. Hence, it will not provide optimised numerical solutions for a process such as this.

Chapter 5   Performance Analysis

5.3 Lax-Wendroff Scheme

This subsection details the numerical solutions, graphical outputs and, the performance analysis of the Lax-Wendroff Finite Difference scheme.

5.3.1 Numerical Solutions and Graphical Outputs

The following highlights the numerical solutions and graphical outputs from the Lax-Wendroff scheme that were obtained for the linear convection of temperature partial differential equation as part of the performance analysis investigation.

5.3.1.1 Numerical Solutions for a Time-Step Size of 0.01s

Table 5.6 shows the Numerical Solutions for the Lax-Wendroff scheme, which were calculated using the Finite Difference Computational Grid System (figure 3.0), for a time-step size of 0.01s ($\Delta t = 0.01$ and consequently $v = 0.05$).

x-direction		n	n+1	n+2	n+3	n+4	n+5	n+6	n+7	n+8	n+9	n+10	n+11	n+12
0	i	0	0	0	0	0	0	0	0	0	0	0	0	0
0.1	i+1	1	0.9	0.803475	0.710704	0.621936	0.537387	0.457244	0.381659	0.310754	0.24462	0.183316	0.126869	0.075278
0.2	i+2	2	1.945	1.883025	1.814695	1.74065	1.661547	1.578053	1.490842	1.40059	1.307968	1.213642	1.118262	1.022466
0.3	i+3	2	2.045	2.082025	2.110816	2.13118	2.142992	2.146189	2.140778	2.126827	2.104469	2.073896	2.035358	1.989161
0.4	i+4	1	1.1	1.199	1.296349	1.391398	1.483507	1.572052	1.656427	1.73605	1.810369	1.878867	1.941065	1.996524
0.5	i+5	0	0.055	0.11495	0.179609	0.248693	0.321874	0.398783	0.479015	0.562127	0.647647	0.735072	0.823874	0.913506
0.6	i+6	0	0	0.003025	0.009317	0.019095	0.032552	0.049852	0.071129	0.096485	0.125986	0.159663	0.197508	0.239478
0.7	i+7	0	0	0	0.000166	0.000677	0.00172	0.003491	0.006192	0.010027	0.015203	0.021926	0.030397	0.040811
0.8	i+8	0	0	0	0	9.15E-06	4.63E-05	0.00014	0.000331	0.000668	0.001211	0.002032	0.003212	0.004841
0.9	i+9	0	0	0	0	0	5.03E-07	3.04E-06	1.07E-05	2.88E-05	6.52E-05	0.000131	0.000241	0.000415
1	i+10	0	0	0	0	0	0	2.77E-08	1.95E-07	7.83E-07	2.36E-06	5.92E-06	1.31E-05	2.62E-05
1.1	i+11	0	0	0	0	0	0	0	1.52E-09	1.22E-08	5.52E-08	1.84E-07	5.08E-07	1.22E-06
1.2	i+12	0	0	0	0	0	0	0	0	8.37E-11	7.55E-10	3.78E-09	1.39E-08	4.17E-08
1.3	i+13	0	0	0	0	0	0	0	0	0	4.61E-12	4.61E-11	2.54E-10	1.01E-09
1.4	i+14	0	0	0	0	0	0	0	0	0	0	2.53E-13	2.79E-12	1.67E-11
1.5	i+15	0	0	0	0	0	0	0	0	0	0	0	1.39E-14	1.67E-13
1.6	i+16	0	0	0	0	0	0	0	0	0	0	0	0	7.66E-16
1.7	i+17	0	0	0	0	0	0	0	0	0	0	0	0	0
1.8	i+18	0	0	0	0	0	0	0	0	0	0	0	0	0
1.9	i+19	0	0	0	0	0	0	0	0	0	0	0	0	0
2	i+20	0	0	0	0	0	0	0	0	0	0	0	0	0
2.1	i+21	0	0	0	0	0	0	0	0	0	0	0	0	0

Table 5.6: The Numerical Solutions for the Lax-Wendroff Finite Difference Scheme when $\Delta t = 0.01$ and $v = 0.05$.

Chapter 5   Performance Analysis

5.3.1.2 Graphical Solutions for a Time-Step Size of 0.01s

Figure 5.6 shows (as a series of graphical outputs) the temperature convection downstream in a fluid flow as time progresses in intervals 0.01s, for the Lax-Wendroff scheme, for a time-step size of 0.01s ($\Delta t = $ **0.01** and consequently $v = $ **0.05**).

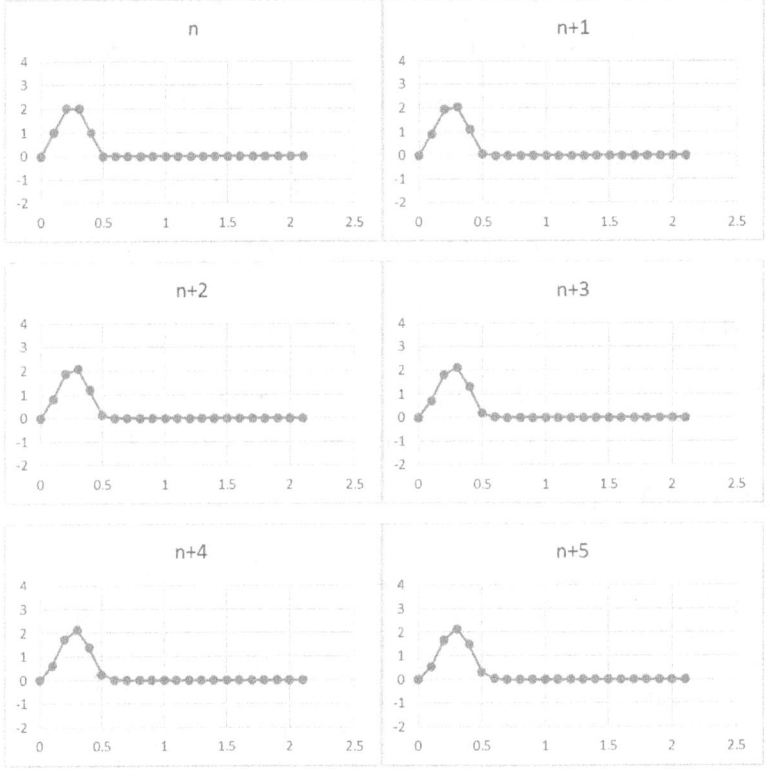

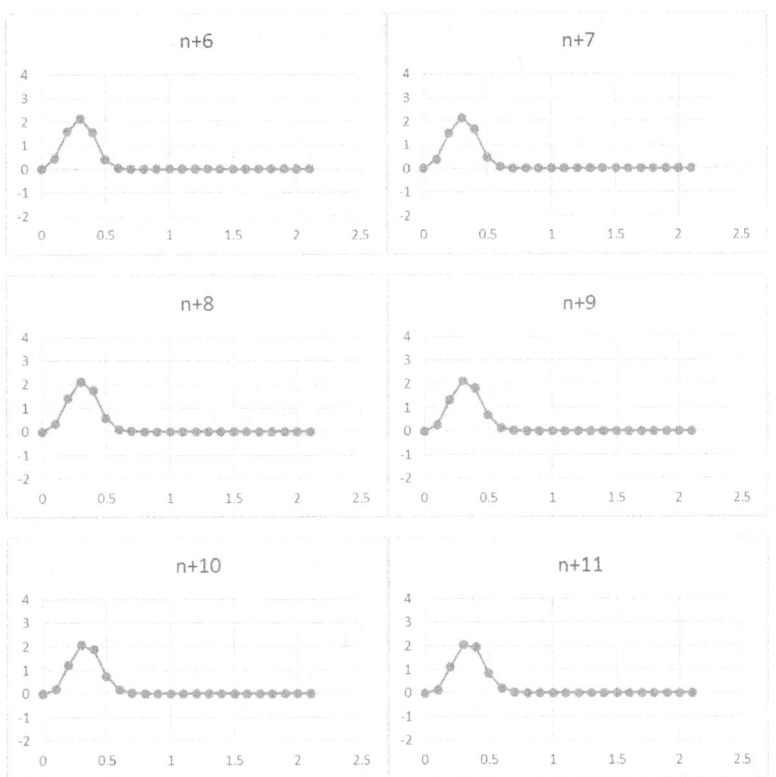

Figure 5.6: Graphs showing the Convection of Temperature Downstream in a Fluid Flow for the Lax-Wendroff Finite Difference Scheme at the 12 Computational Grid Points (Time Intervals of 0.01s) when $\Delta t = 0.01$ and $v = 0.05$.

Chapter 5   Performance Analysis

5.3.1.3 Numerical Solutions for a Time-Step Size of 0.1s

Table 5.7 shows the Numerical Solutions for the Lax-Wendroff scheme, which were calculated using the Finite Difference Computational Grid System (figure 3.0), for a time-step size of 0.1s ($\Delta t = 0.1$ and consequently $v = 0.5$).

x-direction		n	n+1	n+2	n+3	n+4	n+5	n+6	n+7	n+8	n+9	n+10	n+11
0	i	0	0	0	0	0	0	0	0	0	0	0	0
0.1	i+1	1	0	0	0	0	0	0	0	0	0	0	0
0.2	i+2	2	1	0	0	0	0	0	0	0	0	0	0
0.3	i+3	2	2	1	0	0	0	0	0	0	0	0	0
0.4	i+4	1	2	2	1	0	0	0	0	0	0	0	0
0.5	i+5	0	1	2	2	1	0	0	0	0	0	0	0
0.6	i+6	0	0	1	2	2	1	0	0	0	0	0	0
0.7	i+7	0	0	0	1	2	2	1	0	0	0	0	0
0.8	i+8	0	0	0	0	1	2	2	1	0	0	0	0
0.9	i+9	0	0	0	0	0	1	2	2	1	0	0	0
1	i+10	0	0	0	0	0	0	1	2	2	1	0	0
1.1	i+11	0	0	0	0	0	0	0	1	2	2	1	0
1.2	i+12	0	0	0	0	0	0	0	0	1	2	2	1
1.3	i+13	0	0	0	0	0	0	0	0	0	1	2	2
1.4	i+14	0	0	0	0	0	0	0	0	0	0	1	2
1.5	i+15	0	0	0	0	0	0	0	0	0	0	0	1
1.6	i+16	0	0	0	0	0	0	0	0	0	0	0	0
1.7	i+17	0	0	0	0	0	0	0	0	0	0	0	0
1.8	i+18	0	0	0	0	0	0	0	0	0	0	0	0
1.9	i+19	0	0	0	0	0	0	0	0	0	0	0	0
2	i+20	0	0	0	0	0	0	0	0	0	0	0	0
2.1	i+21	0	0	0	0	0	0	0	0	0	0	0	0

Table 5.7: The Numerical Solutions for the Lax Finite Difference Scheme when $\Delta t = 0.1$ and $v = 0.5$.

Chapter 5   Performance Analysis

5.3.1.4 Graphical Solutions for a Time-Step Size of 0.1s

Figure 5.7 shows (as a series of graphical outputs) the temperature convection downstream in a fluid flow as time progresses in intervals 0.1s, for the Lax-Wendroff scheme, for a time-step size of 0.1s ($\Delta t = 0.1$ and consequently $v = 0.5$).

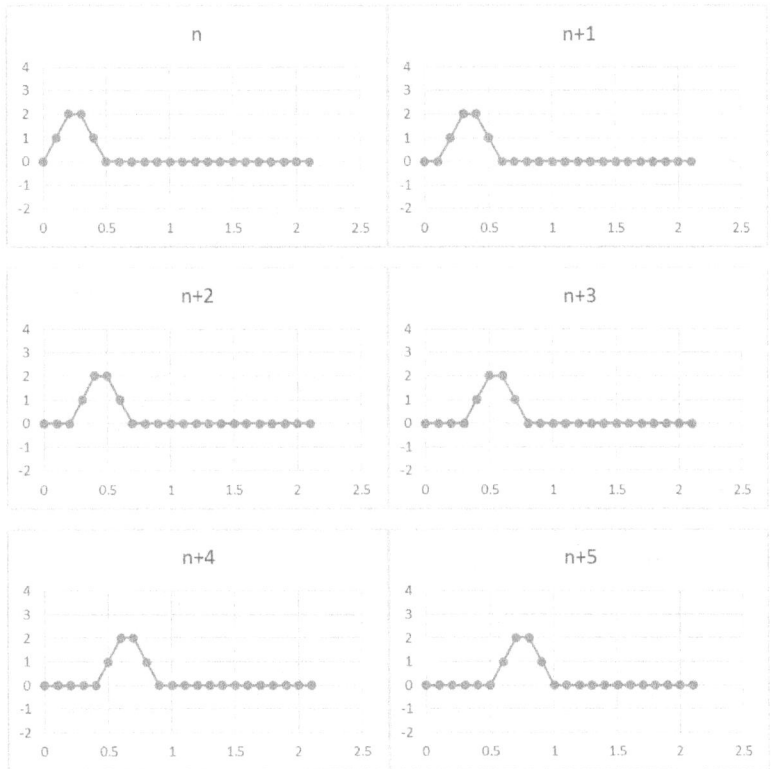

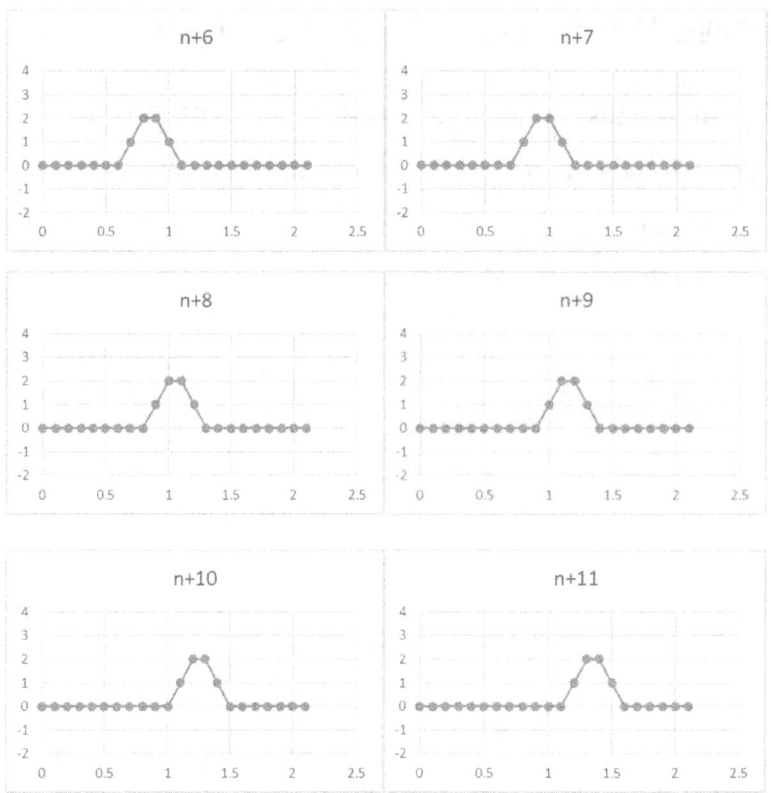

Figure 5.7: Graphs showing the Convection of Temperature Downstream in a Fluid Flow for the Lax-Wendroff Finite Difference Scheme at the 12 Computational Grid Points (Time Intervals of 0.1s) when $\Delta t = 0.1$ and $v = 0.5$.

Chapter 5   Performance Analysis

5.3.1.5 Numerical Solutions for a Time-Step Size of 0.12s

Table 5.8 shows the Numerical Solutions for the Lax-Wendroff scheme, which were calculated using the Finite Difference Computational Grid System (figure 3.0), for a time-step size of 0.12s ($\Delta t = 0.12$ and consequently $v = 0.6$).

x-direction		n	n+1	n+2	n+3	n+4	n+5	n+6	n+7	n+8	n+9	n+10	n+11
0	i	0	0	0	0	0	0	0	0	0	0	0	0
0.1	i+1	1	-0.2	0.1696	-0.11514	0.100165	-0.08974	0.085744	-0.08465	0.086062	-0.0894	0.094518	-0.10138
0.2	i+2	2	0.68	-0.3376	0.412544	-0.38055	0.38549	-0.39106	0.406781	-0.42941	0.459859	-0.49825	0.54537
0.3	i+3	2	1.88	0.3344	-0.39213	0.715236	-0.85826	1.01276	-1.15568	1.310977	-1.4826	1.678129	-1.90333
0.4	i+4	1	2.2	1.672	-0.01549	-0.34352	1.052329	-1.61555	2.212713	-2.82458	3.489755	-4.22712	5.061705
0.5	i+5	0	1.32	2.3232	1.39392	-0.35777	-0.16355	1.375221	-2.71237	4.303759	-6.12163	8.222013	-10.6568
0.6	i+6	0	0	1.7424	2.299968	1.103985	-0.71514	0.210493	1.579475	-4.16279	7.683481	-12.1613	17.73412
0.7	i+7	0	0	0	2.299968	2.023972	0.931027	-1.19333	0.937569	1.424107	-5.83326	12.75084	-22.5271
0.8	i+8	0	0	0	0	3.035958	1.335821	1.12209	-2.06892	2.401828	0.350061	-7.19835	19.65035
0.9	i+9	0	0	0	0	0	4.007464	0	2.115941	-3.94129	5.463191	-2.9003	-6.79521
1	i+10	0	0	0	0	0	0	5.289853	-2.32754	4.65507	-7.9881	11.9207	-11.0396
1.1	i+11	0	0	0	0	0	0	0	6.982606	-6.14469	9.954403	-16.3842	25.33875
1.2	i+12	0	0	0	0	0	0	0	0	9.21704	-12.1665	19.95305	-32.9761
1.3	i+13	0	0	0	0	0	0	0	0	0	12.16649	-21.413	37.68693
1.4	i+14	0	0	0	0	0	0	0	0	0	0	16.05977	-35.3315
1.5	i+15	0	0	0	0	0	0	0	0	0	0	0	21.1989
1.6	i+16	0	0	0	0	0	0	0	0	0	0	0	0
1.7	i+17	0	0	0	0	0	0	0	0	0	0	0	0
1.8	i+18	0	0	0	0	0	0	0	0	0	0	0	0
1.9	i+19	0	0	0	0	0	0	0	0	0	0	0	0
2	i+20	0	0	0	0	0	0	0	0	0	0	0	0
2.1	i+21	0	0	0	0	0	0	0	0	0	0	0	0

Table 5.8: The Numerical Solutions for the Lax-Wendroff Finite Difference Scheme when $\Delta t = 0.12$ and $v = 0.6$.

Chapter 5　　　　　　　　　　　　　　　　　　　Performance Analysis

5.3.1.6 Graphical solutions for a Time-Step Size of 0.12s

Figure 5.8 shows (as a series of graphical outputs) the temperature convection downstream in a fluid flow as time progresses in intervals 0.12s, for the Lax-Wendroff scheme, for a time-step size of 0.12s ($\Delta t = 0.12$ and consequently $v = 0.6$).

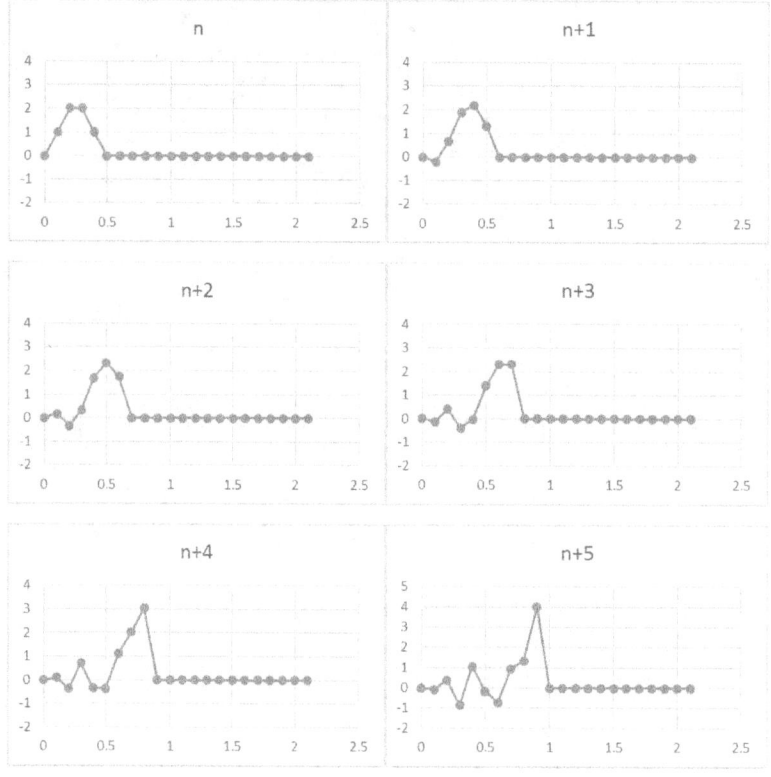

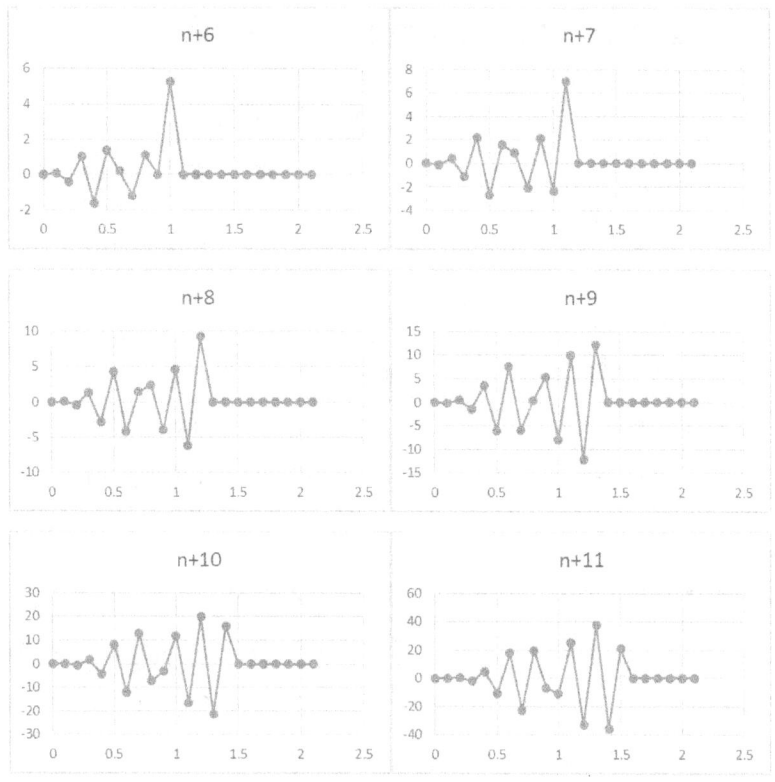

Figure 5.8: Graphs showing the Convection of Temperature Downstream in a Fluid Flow for the Lax-Wendroff Finite Difference Scheme at the 12 Computational Grid Points (Time Intervals of 0.12s) when $\Delta t = 0.12$ and $v = 0.6$.

5.3.2 Lax-Wendroff Scheme Performance Analysis

Similar to the previous schemes (FTBS and LAX), the numerical solutions obtained by using the Lax-Wendroff scheme highlighted numerous trends depending on the size of the time-step employed.

The graphical outputs (figure 5.6) using a time-step size of 0.01s highlight that the Lax-Wendroff scheme is stable, even though the temperature distribution throughout the fluid only underwent minuscule partial change from the initial conditions. This can be attributed to the fact that the time step size used is not large enough to elicit an appropriate response from the Lax-Wendroff scheme, for the numerical solutions of the linear convection of temperature though a fluid. Hence, the convection of the temperature is unable to achieve substantial progress linearly downstream, in this small magnitude of time-period.

This trend is similar to that shown by the results obtained from the FTBS scheme when $\Delta t = 0.01$. Because the temperature distribution does not deviate extensively from the initial conditions, whilst not showing meaningful progression in the x-direction downstream in the fluid flow. As a result, the Lax-Wendroff scheme is inaccurate when such a small time-step is implemented. This extremely limited progression is due to the small amount of time covered by the twelve-time step intervals, 0.12s to be exact. Therefore, the linear convection has been unable to develop effectively.

When comparing the performance of the schemes when stable, it can be observed that the relative performance of the Lax-Wendroff scheme is more comparable to the FTBS scheme. Given that the stable graphical outputs (figure 5.0 and figure 5.6) have similar characteristics, unlike the Lax scheme results (figure 5.3) which appeared to show the temperature distribution dissipating instead of progressing downstream linearly.

As a side-note, a stable Lax scheme does not provide an accurate representation of the model partial differential equation. A predictable trait since the Lax scheme is not convergent. Therefore, the Lax-Wendroff scheme has greater performance accuracy than the Lax scheme when stable. This is expected given that the Lax-Wendroff scheme is consistent with the original partial differential equation, whereas the Lax scheme is inconsistent. The analytical observations of the numerical solutions for $\Delta t = 0.01$, highlight that the relative performance of the Lax-Wendroff Finite Difference scheme is stable but inaccurate for small values of the time-step Δt.

The solutions obtained from the Lax-Wendroff Finite Difference scheme for $\Delta t = 0.1$ elicit the same output response numerically and graphically as the FTBS and Lax schemes. Consequently, this Finite Difference scheme has the same characteristics as those investigated previously when the time-step is 0.1s. The graphical outputs (figure 5.7) across the twelve separate time intervals show the initial temperature conditions moving linearly within the fluid flow in the x-direction, with a constant peak temperature of 2°C. Therefore, accurately providing an exact representation of the linear convection of temperature travelling downstream in a flowing fluid. It can be observed that the graphical outputs do not contain fluctuations or negative values. Hence, when $v = 0.5$ the Lax-Wendroff Finite Difference scheme is stable and accurate.

The Lax-Wendroff scheme being both stable and accurate is to be expected, given that the solutions identically replicate those obtained from the preceding two Finite Difference schemes when the time-step size is equal to 0.1s. Thus, it can be determined that the relative performance of FTBS, Lax and Lax-Wendroff schemes is optimal when $\Delta t = 0.1$s. Since all three schemes provide an accurate and completely stable model of the linear convection of temperature downstream in a fluid flow for this time-step magnitude.

The numerical solutions obtained when $\Delta t = 0.12$, from the Lax-Wendroff scheme, highlight that when the time-step size is 0.12s the scheme produces very inaccurate results for the linear convection of temperature in a fluid flow as time progresses across the twelve various interval points. Additionally, the Finite Difference scheme is considerably unstable with large amounts of fluctuations in the numerical solutions, as displayed by the graphical outputs (figure 5.8).

The amplitude of these fluctuations consistently increases with time, which causes the maximum peak temperature in the fluid to consistently rise as the temperature distribution moves downstream in x-direction. Resulting in excessive inaccuracy of the numerical solutions. This rise in inaccuracy can be attributed to Truncation Errors becoming unbounded. Thus, causing greater inaccuracies at each time interval, which has a considerably larger deterioration impact on the subsequent numerical solutions at following time intervals downstream in the fluid flow. This is exacerbated when the Lax-Wendroff Finite Difference scheme is unstable.

A comparative evaluation of the effects of Truncation Errors on numerical solutions obtained from the FTBS and Lax schemes, shows that Truncation Errors are a detrimental performance flaw of the Lax-Wendroff Finite Difference scheme.

Overall, the Lax-Wendroff Finite Difference scheme is consistent with the original partial differential equation as the time step size tends to zero. However, it becomes progressively unstable and inaccurate for a time-step size of 0.12s, as verified by thorough analysis of the graphical outputs results (figure 5.8). Hence, the scheme becomes more unstable and inaccurate as time-step size is increased further. Moreover, its performance is sub-optimal when the time-step size, exceeds the limit of stability (i.e. is greater than 0.1s). Furthermore, its performance will be significantly diminished as time-step size increases above this limit.

Consequently, for optimal performance the Lax-Wendroff Finite Difference scheme should operate with a time-step size of 0.1s when employed to model a process that is described by linear partial differential equations.

Discussion

This chapter of the book provides a discussion of the key performance areas for the Finite Difference method and its associated schemes, when used to provide numerical solutions for partial differential equations. Encompassing, the effects of scheme Consistency, Stability, Truncation Errors and the impact of time-step size on the performance capabilities of the FTBS, Lax and Lax-Wendroff, Finite Difference Schemes.

Furthermore, an evaluation of the three schemes' performance during the performance analysis investigation is present in this chapter. Culminating in a recommendation of the optimal Finite Difference scheme to use to obtain the most accurate and stable numerical solutions for processes which are modelled by linear partial differential equations.

6.1 Importance of Consistency

The consistency of a Finite Difference scheme is a crucial measure of its ability to approximate accurate numerical solutions for original partial differential equations. Therefore, consistency is critical to the accuracy and performance of a Finite Difference scheme (Frey, 2009)[14]. A Finite Difference scheme is defined as consistent if, the Taylor Series expansion of the schemes discretised equation reduces to the form of the original partial differential equation being modelled when the independent variables tend to zero (Urroz, 2004)[12]. Hence, the consistency of a Finite Difference scheme is dependent on the Truncation Error becoming zero as Δt and Δx tend to zero.

Nevertheless, the implication that a consistent Finite Difference scheme is always useful is incorrect, given that convergence is the true measure of accuracy. Convergence is dependent on the stability of the scheme. The consistency investigation undertaken (Taylor Series Expansions chapter) highlighted that the FTBS and Lax-Wendroff schemes were consistent, but the Lax scheme was inconsistent. Thus, inferring that the relative performance of the Lax Finite Difference scheme, when modelling the linear convection of temperature, is inadequate because the scheme is inconsistent. Hence, it is not convergent, because the discretised Lax scheme is unable to replicate the original partial differential equation (equation 2.0).

Consequently, Finite Difference scheme consistency is critically important, because the scheme requires 100% consistency with the original partial differential equation to be convergent, as determined by the Lax Equivalence Theorem, to be accurate. Without consistency a scheme can never produce wholly accurate, nor reliable numerical solutions for the dynamic process or phenomena that is being modelled. Due to the constant presence of Truncation Errors in the solutions.

6.2 Effects of Truncation Error

Truncation Error has a considerable effect on the relative performance of a Finite Difference scheme because it negatively effects both the consistency and stability. If the Truncation Error in a scheme does not tend to zero as the spatial and time-step variables tend to zero, then the scheme will be inconsistent. Hence, it will have poor performance when modelling a partial differential equation. Subsequently, if the Truncation Error in the numerical solutions is too large it will result in the scheme being unstable because the

Truncation Errors will continuously grow as time progresses. Thus, becoming unbounded between continuous solutions. Causing the scheme to have a poor relative performance.

This phenomenon can be observed from the fluctuating numerical solutions and graphical outputs of the three schemes when they are unstable (figures 5.2, 5.5 and, 5.8). As they show the Truncation Errors in the results increasing at each successive time interval. Consequently, increasing instability. Although Truncation Errors can compound instability of a Finite Difference scheme, by deteriorating performance and the accuracy of numerical solutions, it is not the dominant cause of a scheme to be unstable. Since, the stability and size of Truncation Errors are dependent on the time-step size.

6.3 Impact of Time-Step Size on Scheme Stability and Performance

The size of the time-step used in a Finite Difference scheme is the most influential variable on its stability, given that if too large a time-step is used the numerical solutions and, the Finite Difference scheme will become unstable.

This is shown by the results that were obtained from the performance analysis investigation of varying the time-step to determine the performance change of each scheme. The graphical outputs (figures 5.0 - 5.8) highlight that the time-step size has a considerable impact on the performance, accuracy and stability of each of the three schemes. It has been established that a large time-step size ($\Delta t = 0.12$) that is greater than the time-step limit of stability ($\Delta t = 0.1$) will produce inaccurate and diverging numerical solutions. Solutions that contain continually growing unbounded errors.

This is due to instability in the Finite Difference schemes. As shown when a large time-step size of $\Delta t = 0.12$ caused numerous fluctuations in the numerical solutions, for the FTBS, Lax and Lax-Wendroff schemes (figures 5.2, 5.5 and, 5.8). This growing instability, due to Truncation Errors mainly, instigates progressively increased fluctuations in the frequency and amplitude of the temperature distribution at each consecutive time interval as it travels downstream in the fluid.

The scheme that was most susceptible to instability and inaccuracy when $\Delta t = 0.12$ was the Lax-Wendroff. Therefore, the gradient of the performance degradation curve for the Lax-Wendroff scheme is substantially higher than its counterparts (FTBS and Lax) as the time-step size increases. So, its performance will suffer greater as Δt increases.

In contrast when a very small-time step is implemented, the schemes are stable but produce inaccurate results because the temperature distribution stayed relatively stationary. Thus, not showing progression linearly in the x-direction. Hence, the performance characteristics of the schemes are diminished when an excessively small-time step size is employed. Moreover, the Lax scheme was the most inaccurate when a small-time step size was used, showing the temperature dissipating instead of moving in the x-direction. This can be attributed to the scheme being subject to excessive damping. Moreover, it is inconsistent with the original partial differential equation when Δt and Δx tend to zero, which further contributes to poor performance.

6.4 Discussion of Finite Difference Method Scheme Performance

The relative performance of the three Finite Difference schemes is primarily dependant on the size of the time-step. The FTBS, Lax and Lax-Wendroff schemes are all unstable when the time-step size exceeded the limit of stability, $\Delta t = 0.1$. As expected, given the major impact stability has on numerical solution accuracy. The FTBS and Lax schemes have similar results and levels of instability. Therefore, are comparable in accuracy when unstable.

Whereas the Lax-Wendroff scheme is considerably more unstable and inaccurate than the other schemes when the time step exceeded the limit of stability, $\Delta t > 0.1$. Since it produces highly divergent fluctuating results that progressively worsen at each time interval. This is due to the discretised equation for the Lax-Wendroff scheme (equation 4.16) containing more temperature variables that are subject to Truncation Errors when the scheme is unstable. Hence, the growth of the Truncation Errors in the Lax-Wendroff scheme will be larger because there is more terms that are subjected to Truncation Errors in this scheme than the FTBS and Lax. As a result the performance of the Lax-Wendroff scheme when it is unstable is inferior to that of the FTBS and Lax schemes.

When the time step size was 0.1s all three schemes have the same level of performance because they produce the same accurate results and are stable showing the initial temperature profile progressing linearly (as shown in figures 5.1, 5.4, and 5.7). Therefore, accurately modelling the linear convection of temperature in a fluid flow. Nonetheless the Lax scheme is inconsistent. Hence, it is not convergent, so its performance is still questionable and should not be employed to accurately model a process that is described by a linear partial differential equation.

6.5 Evaluation of Scheme Performance

Analysing the results of the stability and consistency investigations enables the relative performance of the three different schemes when applied to the linear convection equation to be evaluated.

The relative performance of the FTBS scheme varies depending on the size of the time-step, but overall its performance is satisfactory and consists of many advantages. Such as; it is less effected by Truncation Errors than the Lax-Wendroff when unstable, it is consistent, and can accurately model the linear convection of temperature. The FTBS scheme does not have any real disadvantages other than its simplicity, which will inhibit its performance for more complex partial differential equations and when employed to model multifaceted processes.

The investigation highlighted that the Lax scheme has inadequate performance when applied to the linear convection equation (equation 2.0), because of its disadvantageous characteristics of inconsistency (it is not convergent) and excessive damping of stable solutions. Although it does provide the most effective performance when unstable.

Finally, the investigation shows that the Lax-Wendroff scheme has similar performance attributes to the FTBS scheme. Hence, it has the same advantages. However, its performance is severely hampered by unique disadvantages such as; excessive unbounded fluctuations and extensive Truncation Errors when unstable. Therefore, effective relative performance can only be achieved when the scheme is stable.

Chapter 6   Discussion

6.6 The Optimally Performing Finite Difference Scheme

By completing the performance analysis investigation and analysing the results obtained, it can be determined that the optimum Finite Difference scheme to use for the most accurate numerical solutions for the linear convection of temperature equation is the Forward-Time, Backward-Space scheme. Given that, it is more accurate than the Lax-Wendroff scheme when unstable and is consistent unlike the Lax scheme. As a result of the Lax scheme being inconsistent with the original partial differential equation, it was determined to have the worst performance, irrespective of it being the most accurate when unstable.

Investigation Conclusion

Overall the investigation into the relative performance of the FTBS, Lax and Lax-Wendroff Finite Difference Schemes was a successful exercise, given that all the aims and objectives were fulfilled. The investigation established that the three Finite Difference schemes were all stable for time-step sizes less than 0.1s and unstable when larger time-step sizes, that exceeded the limit of stability ($\Delta t > 0.1$) were implemented. Hence, the investigation proved that time-step size has an extreme influence on scheme stability.

Furthermore, the investigation established that the relative performance of the schemes when applied to the linear convection equation (equation 2.0) varies depending on the size of the time-step and the consistency of the Finite Difference scheme. The Numerical solutions and graphical outputs (figures 5.2, 5.5 and, 5.8) highlight that the Lax scheme has the most effective performance when unstable, given that its results are most accurate and have less fluctuations, when stability is compromised. The accuracy of results and performance of the FTBS and Lax-Wendroff schemes is similar and much better than that of the Lax scheme when stable.

After completing the performance analysis investigation and analysing the accuracy of the numerical solutions it was determined that the most suitable scheme of the three investigated to model the linear convection of temperature in a fluid flow is the Forward-Time, Backward-Space Finite Difference scheme. Firstly, it is consistent with the original partial differential equation when Δt and Δx tend to zero. Whereas, the Lax scheme is inconsistent. Therefore, when the FTBS scheme is stable it will be convergent. Thus, enabling accurate results to be obtained.

Furthermore, the FTBS and Lax-Wendroff schemes are both convergent, in addition to having similar accuracy and performance characteristics when stable. Yet, when they are unstable the FTBS is more accurate. Because, the divergence and Truncation Error in the results is less excessive than that observed in the Lax-Wendroff results. This is due to the Lax-Wendroff scheme being subject to increased Truncation Errors. Hence, the FTBS scheme performs better when larger time-step sizes are implemented, and it is unstable. Resulting in it being more suitable than the Lax-Wendroff scheme to model and provide solutions for processes that are modelled by linear partial differential equations.

7.1 Sequential Investigation Suggestions

Having established a definitive outcome of the performance analysis investigation, that being the FTBS Finite Difference scheme is the most suitable to model and provide solutions for linear partial differential equation, it is appropriate to provide suggestions for additional investigations that build on the concepts of this book and findings of the performance investigation. A subsequent investigation should conduct additional research into the effects of a non-constant convection or intermittent fluid flow velocity, to observe how this affects overall scheme performance, stability, convergence, and numerical solution accuracy.

7.2 Improvements to the Investigation

Improvements to the numerical study could be achieved by using a finer grid mesh (i.e. a smaller Δx) to determine how the relative performance of the schemes changes. Additionally, a more varied set of stable values of Δt, which fall below the limit of stability, $\Delta t < 0.1$, should be investigated and used to calculate numerical solutions. This will provide comprehensive understanding of the influence of time step size on the accuracy of numerical solutions obtained from stable, consistent and convergent Finite Difference schemes.

Although there are numerous options for further research and possible refinements to improve the Finite Difference scheme performance investigation, this investigation was effective. Since, it enabled the relative performance of the schemes to be determined as well as providing suitable results for the linear convection of temperature in a fluid flow.

References

1. Thomee, V., 2001. From Finite Differences to Finite Elements. [Online]
Available at: http://www.math.chalmers.se/Math/Research/Preprints/1999/21.pdf
[Accessed 15 12 2015].

2. Christian Grossmann, H.-G. R. M. S., 2007. Numerical Treatment of Partial Differential Equations. Berlin : Springer-Verlag Berlin Heidelberg.

3. Bernoff, A. J., 2008. An Introduction to Partial Differential Equations in the Undergraduate Curriculum. [Online]
Available at: https://www.math.hmc.edu/~ajb/PCMI/lecture1.pdf
[Accessed 02 12 2015].

4. Professor D.M. Causon, P. C. G. M., 2010. Introductory Finite Difference Methods for PDE's. s.l.:Ventus Publishing ApS.

5. University of Alaska, F., 2004. Chapter 9: Convection Equations. [Online]
Available at: http://how.gi.alaska.edu/ao/sim/chapters/chap9.pdf
[Accessed 14 12 2015].

6. University, Y., 2009. Chapter 5: Finite Difference Methods. [Online]

Available at: http://www.math.yorku.ca/~hmzhu/Math-6911/lectures/Lecture5/5_BlkSch_FDM.pdf
[Accessed 16 12 2015].

7. Johnston, D. L., 2014. Finite-Difference Approximation. In: Aerodynamics M1: Lecture Notes. Salford: University of Salford, p. 9/61.

8. MIT, 2010. Finite Difference Methods. [Online]
Available at: http://web.mit.edu/course/16/16.90/BackUp/www/pdfs/Chapter13.pdf
[Accessed 15 12 2015].

9. Trefethen, 1994. Accuracy, Stability and Convergence. [Online]
Available at: https://people.maths.ox.ac.uk/trefethen/4all.pdf
[Accessed 15 12 2015].

10. Stanford, U. o., 2008. Convergence, Consistency and Stability. [Online]
Available at: http://web.stanford.edu/class/cme306/Discussion/Discussion1.pdf
[Accessed 14 12 2015].

11. J.H. Ferziger, M., 2002. 2.5.3 Convergence. In: Computational Methods for Fluid Dynamics. Berlin: Springer, p. 46.

12. Urroz, G. E., 2004. Convergence, Stability, and Consistency of Finite Difference Schemes in the Solution of Partial Differential Equations. [Online]

Chapter 8   References

Available at: http://ocw.usu.edu/Civil_and_Environmental_Engineering/Numerical_Methods_in_Civil_Engineering/StabilityNumericalSchemes.pdf [Accessed 03 12 2015].

13.   Hirsch, C., 2007. 7.1.1.1 Methodology. In: Numerical Computation of Internal and External Flows . Oxford : Elsevier, pp. 288-291.

14.   Frey, C., 2009. The Finite Difference Method. [Online] Available at: http://www.ann.jussieu.fr/~frey/cours/UdC/ma691/ma691_ch6.pdf

www.ingramcontent.com/pod-product-compliance
Lightning Source LLC
Chambersburg PA
CBHW062355220526
45472CB00008B/1819